KEITH BROCKIE'S
Wildlife Sketchbook

Macmillan Publishing Co., Inc.

New York

Macmillan Publishing Co., Inc.
866 Third Avenue, New York, N.Y. 10022
Collier Macmillan Canada, Ltd.

First American Edition 1981

Printed in Italy by
Arnoldo Mondadori Editore, Milan

Library of Congress Catalog Card Number 81-3769
ISBN 0-02-516450-3

Contents

Mountains and Moorland

I cannot imagine a better combination for a satisfying life than a talent to draw and an interest in nature. Keith Brockie's book gives me a quite un-Christian feeling of envy. I would love to have his talent and the time to indulge it.

The drawings and notes also betray a keen eye and a sympathetic understanding for the problems of wildlife. Altogether it is a most attractive reminder of the richness and beauty of nature in Scotland.

1981

Introduction

'When did you start drawing?' is a question people often ask me
when they see my work. As to the answer, I think I have always
drawn since childhood, encouraged both by my parents and my
teachers. Neither of my parents and none of my relatives ever
really drew or painted, so I didn't inherit any ability to draw.
Undoubtedly the move from Haddington, East Lothian to
Strathmiglo, Fife in 1968, when I was thirteen years old,
strengthened my interest in wildlife, because the marvellous
wildfowl populations on Loch Leven and the Highlands were
now not too far distant. And yearly holidays to Little Loch
Broom in Wester Ross, where I could explore the rugged hills
and coastline, further kindled my interest in highland wildlife.

The move to Bell Baxter High School in Cupar, Fife, played
an important part in my artistic development because there was
a particularly good art department there. Two teachers, Bob
Crerar and Will Maclean, really helped to cultivate my interest in
both wildlife and art at that impressionable age when the mind
easily wanders and one tends to change interests quickly. Their
intense concern for art outside school life was especially
inspiring. Bob's great loves were wildlife, fossils, archaeology
and jewellery, and Will was a devotee not only of wildlife, too,
but of Scottish west coast fishing traditions. Learning from them
and talking with them was a marvellous experience. Following
school I spent four years at Duncan of Jordanstone College of
Art which gave me a general grounding in most forms of art and
design and later I specialized in illustration and printmaking
under the guidance of Ron Stenberg. Here I had a chance to
learn a great deal from the work of other students and lecturers.
After college I spent a few months working as an illustrator for
Dundee Museum but gave this up in June 1979 to work as a
freelance artist.

Some bird artists like to draw entirely from specimens in
museums. They are very good at it, too, but it is not my style. I

find it much more refreshing to work and draw in the field, in the wild itself. Seeing the work and sketches of highly individual artists such as John Busby, Donald Watson and Doug Weir, all resident in Scotland, provided me with an early insight into the ways of depicting wildlife as it really is. I recently spent a week with John Busby at a course run by Dr Eric Ennion for wildlife artists. It was a privilege to see his vast array of work done over many years. His unique style of depicting wildlife, especially birds, on the move, with such economy of line and colour, really brings them to life. It is this sort of wildlife art which I find most impressive.

Living in or near the Scottish countryside from an early age meant that wildlife, and birds in particular, were always destined to be my great passion. I think it was inevitable that they should feature prominently in my drawing. Many young people who become interested in the natural world around them are unwittingly apt to commit cardinal sins, such as egg-collecting. But most birdwatchers I know have quickly put such indiscretions behind them and progressed from there to watching the birds in order to find out more about their habits and identification. Gradually I began to take just as much interest in the landscape around the birds, learning about plants and animals as well. It is only in the last few years, however, that I have put a lot more effort into drawing instead of just watching wildlife.

Like most other professions, probably more than most, wildlife art is a continual process of learning and the importance of fieldwork means that you discover new things every day, thus helping you maintain a fresh approach to the ways of depicting wildlife. For instance, checking species like herons and peregrine falcons means that little if any drawing can be done on site, because disturbance must be kept down to the minimum level possible. In any case the time involved in scaling trees and cliffs often precludes any drawing at all. But the knowledge and experience gained from exploring such terrain pays great dividends in the end.

For anyone interested in wildlife, Scotland is a remarkable country. Because it is so small an area it is relatively quick to get away from urban life into the isolation and wild beauty of the countryside, where there is so much to explore, and draw, that it is hard to know where to start. With so many rich and varied habitats for its wildlife, Scotland is an artist's paradise. With its islands like St Kilda, Skye and Orkney, its rugged coastlines, its rivers and lochs that lead into the forests, wild glens and moorlands that in turn stretch out below the high tops of the mountains, there is no question that Scotland is one of the last great wildernesses of Europe.

Standing among the remains of the old Caledonian pines in the Black Wood of Rannoch (Perthshire), Rothiemurchus (Speyside) and Mar (Deeside), one can get an impression of what Scotland once used to look like. In the sixteenth century most of Scotland was carpeted with forests of Scots pine, oak, ash, hazel and aspen, with a scrub layer of birch, willow and juniper above. Wolf and wild boar used to run free, beavers swam in the rivers and streams, and ospreys, red kites and goshawk all nested freely. But it was from the sixteenth century onwards that alterations to the landscape in Scotland, as elsewhere in Britain, became increasingly common; and not the least of the forms of man's exploitation was the introduction of sheep, whose grazing prevented regeneration. In fact, the bare glens and moorland now over much of Scotland, which so attract the visitor, are really an unnatural habitat for their wildlife.

Nowadays new forests are being planted where the old ones used to grow. Monoculture forests of foreign pines such as the sitka spruce and lodgepole pine are extending back up onto the hills. Economic considerations dictate the large-scale planting of vast acres of trees. These forests may be good for wildlife in their early years with small trees and undergrowth of grass but they support far fewer species when the trees grow tall and their closed canopies prevent any scrub layer. Perhaps one solution would be for forestry operations to allow more diverse planting in the future when these forests become established and wood is felled in rotation. On the plus side, though, it has to be said that mammals like the fox, wildcat and pine marten are benefiting from afforestation and extending their ranges.

Many of Scotland's unspoiled habitats are at risk, and there are two dangers in particular. First is the ever-increasing possibility of a major oil spill which could devastate the sea-bird populations, such as the gannets, for which Scotland is internationally important. Thousands of sea-birds already die every year from minor oil spills around our coast but the thought of a major disaster during the summer off the sea-bird

cliffs defies the imagination. The insidious pollution of the shoreline by chemical residues is being increasingly detected, though the consequences are still largely unknown. In the past, certain species of bird have been extinguished in Britain by man's actions. Any wildlife lover has to ask himself which will be the next creatures to be doomed by today's industrial needs?

Secondly, the increased mobility enjoyed by many today has led to much more pressure on the fragile ecology of some of the mountain areas of Scotland. Skiing in particular is expanding every year with developments in the Cairngorm and Cairnwell regions causing great friction between conservationists and the chairlift companies. At issue is the erosion caused by the initial development and then the increased pressure on the terrain by the numbers of people brought up by the chairlift onto the plateaux during the summer. Unwittingly, they disturb the rare wildlife, besides strewing litter over the tops. I am sure this expansion could be made in less sensitive areas.

Tracking down and observing wildlife is a challenge in itself even if I had no aspirations to draw the species I was searching for. Plants are sometimes difficult to locate but of course, being stationary, pose no real problems when you find them. Birds and animals are a different matter. They are often just as difficult to locate, and frequently on the move. For instance, it is extremely difficult to find a species such as the ptarmigan, especially on its nest, not only because of its cryptic camouflage but also because it lives so high up in the mountains. Throughout 1980 I searched for ptarmigan for many hours, sometimes with friends, and found quite a few males on sentry duty, pairs not yet nesting and newly fledged young. But we didn't come across any females actually nesting. Then one day on the summit of Driesh, Glen Clova, I thought we had finally succeeded when one of my friends spotted a female sitting with a male in attendance. Alas, she got up and walked away before I could put pencil to paper – anyway, she was sitting on an empty scrape! However, we did find other forms of wildlife on these searches, such as the tiny mountain hare leveret, a just recompense for our efforts.

Hides are of great benefit for observing wildlife, especially permanent hides overlooking lochs or estuaries. Portable hides

set up at nests give me a splendid opportunity to observe a bird's nesting habits, for instance watching a female merlin trying to shield two large chicks for hours in pouring rain, gradually becoming more bedraggled and uncomfortable. At an Arctic skua's nest in Orkney I needed the hide more for protection than anything else, for not only did both adults swoop down and try to strike me on the head as I approached the chicks, they actually landed on my back as I bent down to examine the chicks.

Portable hides have two big drawbacks in that firstly I need another person to walk with me to the hide, if it is at a nest, and walk away leaving me in the hide. The birds then think all is well and come back to the nest. But also, before I can even begin sketching, the hide has to be moved in stages nearer to the nest which takes up precious time. A car is a good hide for drawing birds and animals from the roadside, such as the Bewick's and whooper swans I drew near Loch Leven, Kinross-shire. Whether I'm in a hide, a car or out in the open, I always use a telescope with zoom magnification for drawing animals and birds. Mine is a 25 to 60× magnification mounted on an adjustable tripod which I can use either while standing or sitting. This leaves my hands free to draw while still watching the animal or bird; it is so much better than having to pick up binoculars all the time, with the consequent loss of time and concentration.

For many species I need in advance either a rough knowledge of their habitat preferences and behaviour patterns in order to locate them, or else a knowledgeable person along with me to advise me. Although books can teach one a lot about individual species there is no substitute for the personal experience and help of the people who are working with particular species in the field. Scientists, reserve wardens, fellow ringers, gamekeepers and stalkers, to name but a few, have taught me a lot about the species they are studying. More often than not they even let me accompany them, provided I don't interfere with their work. For instance I learnt more in a few weeks about red deer from stalker Bert Hardie and his ghillie Stuart Rae than I ever knew before. Accompanying them on the hill with their clients to shoot stags in September and October, and later whilst culling the hinds from late October into January, was a magnificent insight into the ways of deer, even though conditions were not often conducive to drawing. How different the scene in September and October from the peaceful months of the rest of the year. Stags challenge each other, roaring night and day, the bellowing and the clash of antlers resounding in the glens. Large master stags, black from rolling in peat hags, guard their groups of hinds, then later when the stags and hinds have separated into different herds, the culling of the hinds begins amidst the backdrop of the first snowfalls. It is a spectacular sequence of events. But at the same time I was careful not to miss the abundant wildlife all around me, from the golden eagle, ptarmigan and grouse to the small parties of snow buntings – all providing me with information to store away for future use.

I had similar, invaluable help in Orkney from Nick Picozzi, a scientist with the Institute for Terrestrial Ecology, who taught me a lot about the wildlife on the island, especially the hen harrier and that dashing little falcon, the merlin. Nick has been studying the hen harriers for many years on the Orkney moorland, monitoring their breeding cycle and polygyny – some males with up to four females each. He also showed me kestrels nesting in the heather and a raven's nest on a small heather bank. Without his help, I would have had to spend a lot of time searching unfamiliar habitats and would have done less than half the drawings that I actually did on Orkney.

Being a bird ringer has helped me immensely in my drawing, enabling me to handle many species of wild bird. Whoever said 'A bird in the hand is worth two in the bush' certainly knew what he was talking about – at least as far as wildlife art is concerned. The ringing scheme in Britain is administered by the British Trust for Ornithology which co-ordinates the work done by amateur and professional ornithologists. One can only gain a permit after special training in all aspects of ringing. There are basically fifteen sizes of rings with different variations to fit all the species of bird in Britain from the tiny goldcrest to the swans and eagle. The rings are put on the birds' legs with special pliers either when they are chicks or fully grown birds, caught by nets and various traps, none of which harm the birds. Although only a small proportion of ringed birds are found again the results yield much valuable information. Their lifespan (over thirty years for some species), migration routes, wintering and summering areas, and seasonal mortality are but a few examples of the invaluable data which ringing can yield. I began ringing with the Tay Ringing Group in 1972, trained mainly by David Oliver, and since then have ringed many thousands of birds of over 160 species. My special interests are in monitoring and ringing herons, birds of prey and waders. To date my best ringing recovery was a heron chick which I ringed in Tentsmuir, Fife, in May 1975 and which was shot at Kasba Tadla, Morocco, seven months later – the first British ringed heron to reach Africa.

Apart from the scientific data gained from ringing birds, the regular practice of handling them gives one the knowledge to determine their sex and age, to differentiate between their different plumages, colours of eye, bills and legs, and generally to get the 'feel' of them. I have found, too, that ringing has given me more of an interest in the 'commonplace' birds, whereas the current trend of most birdwatchers is to amass yearly tick-lists of rare birds.

From ringing I also learn more about the birds' life cycle throughout the year, whether by day or night. For example, the Tay Ringing Group has for many years now been working on the purple sandpiper, a small wader which frequents the rocky shoreline of Scotland during the winter months. In the summers of 1978 and 1980 I took part in expeditions to the Hardangervidda Plateau in the mountains of Norway to follow the purple sandpipers to their breeding grounds on the tundra. There we observed the breeding cycle of these birds and put individual combinations of plastic colour rings on the chicks and adults, hoping to find them on the Scottish coastline in the subsequent winter months. To date three of these birds have been seen on our coastline, one near Aberdeen for the last three winters. Such a complete picture of a bird's yearly life cycle undoubtedly helps me to draw them. I can almost reach into the soul of the bird and, in this way, feel and think like the creature itself.

Drawing birds poses an infinite variety of problems. As I've indicated they are almost always on the move. The colours and plumage vary from bird to bird of the same species, and different lights throughout the day alter this colouring. There is no such thing as a typical textbook bird and individuals vary as much as humans do. By constant observation one has to learn their various plumages – male, female, adult, immature, juvenile, summer, winter and eclipse, for example. Birds will

look very different in the spring compared with the end of the summer before they moult their faded and worn plumage. Some species have more than one moult a year – adult ducks change their body plumage three times during the year. Different geographical variations in some species, such as the different races of the dunlin, further complicate matters. A proper guide to birds depicting all these points would probably be a lifetime's work and would need at least four pages of illustrations per species. Someday I hope to follow just one species throughout the year, producing a monograph in sketches, from which I think I would learn much that would in turn relate to many other species.

Drawing from injured or tame birds and animals helps me to get to know an individual creature, for I can sketch it at leisure, learning how, in a bird's case, it preens, bathes, stretches its wings, feeds, walks and roosts. With certain animals, however, such as deer, foxes, wildcats and badgers, the only way to avoid spending too many frustrating hours in the field, often without any results, is to visit a wildlife park, or zoo, or a deer farm. I also keep a collection of injured wild ducks and geese at home, and people bring me other birds such as tawny owls and buzzards, which I keep as long as they are not in too bad a condition. Many of them I later repatriate back into the wild.

Dead birds, if fresh, provide ideal subjects for drawing. Years ago I saw some reproductions in a magazine of post-mortem drawings done by the late Charles Tunnicliffe. I was very impressed by them. Many museums now have bird skin collections which are available for reference but I have usually found that they are really only of use for plumage colouring. The proportions of the bird are not true to life because of insufficient (or additional) padding by the taxidermist, and also because of natural shrinkage, and the closed wings prevent one learning any details of outstretched wings. In addition the bare parts such as the bill and legs have usually shrivelled and lost their true colours.

So drawing a dead bird is only of real value if it has recently died and is still in good condition. Details of the bird's proportions, its outstretched wings, its bill, gape and feet can then be drawn to scale. Care has to be taken, however, as the colouring of the bill, eyes and legs often fades very quickly after death. I have been fortunate this year to be able to do post-mortem drawings of what I call priority species, that is birds which I am not likely to get the chance to draw in a fresh condition for a long time, such as the storm and Leach's petrel, the sea eagle, goshawk and black-throated diver. Getting to know the birds in this way enables me to delineate the species at a later date in the field with more speed and confidence, knowing that, if needed, I can later fill in details of plumage from my earlier, accurate drawings. I also keep a collection of dried, outstretched wings sealed in clear plastic bags for quick reference.

I also do reference drawings for the mammals which I either find dead or am given by someone else. For example, in sketching the proportions of the bodies of mountain hares and otters and the structure of their heads and feet, I discover that the hare's oversize rear feet are heavily furred to act as snowshoes in the winter snows, and that an animal could hardly be as well adapted for its environment as the aquatic otter, built purely for its speed and agility under the water.

My two favourite mammals are first and foremost the lovely roe deer and second the mountain or blue hare. The roe is a sleek and very graceful deer with a beautiful head, especially the buck with his small pearled antlers. Their coats vary from the foxy-red summer coat to the dark browns of the winter months, not forgetting the dappled coats of the young fawns. Meanwhile the mountain or blue hare of the high tops and moorland is well camouflaged with its thick white winter coat against predation by the golden eagle and hill fox, and has ample protection against the harsh winter snows. Tiny leverets can be found concealed amongst the sparse vegetation on the boulder fields. The soft browns and bluish flanks of its summer and autumn coats conceal the hare until one fortuitously flushes it out, then it bounds away with an easy stride before stopping to look back. Its appearance and characteristics make it a most attractive creature to draw.

When drawing in the field it is not practical to carry a lot of equipment especially if I am walking a long distance. Normally I carry most of my gear in a game bag, including a small drawing board, a sketchbook, some artist's board, a selection of ordinary pencils, watercolour pencils, conté, ball point and felt pens, a half-pan watercolour box and brushes. Together with binoculars, telescope, tripod and waterproofs this is usually more than enough to carry around. Early on in the breeding season, when one has to climb high up cliffs and trees, I usually use a large rucksack as the drawing gear has to compete for space with climbing equipment such as ropes, helmets and climbing irons.

Working on a sketchbook of this nature involved a lot of travelling, getting from place to place on foot or by boat and car. This was very time-consuming compared with working in a small area which I know well because a lot of time was taken up exploring unfamiliar surroundings for suitable subject material. For example, getting to and from St Kilda involved more than 600 miles by car, 220 by bus and hitch-hiking and 44 hours sailing by ferry and landing craft. On the return journey I had to go via Rhu on the Clyde and get back up to Uig in Skye to collect my car before heading home.

Inevitably in Scotland the weather is one of the biggest drawbacks when working in the field, and 1980 was one of the worst on record. For example, I tried for ages to get onto the Isle of May in the Firth of Forth to sketch some grey seal pups in November. The wind never let up for long and when it did the swell would have prevented a landing. Whenever I tried to draw mountains low cloud was bound to be covering the tops, or it was so bitterly cold that my hands were numb. During the few days in Skye before crossing to St Kilda via the Uists I was going to do some landscape drawing in the Cuillins but I couldn't even see them because of low cloud, and since there wasn't a breath of wind for two days the hordes of midges were unbearable. However, it did clear long enough for me to get a sketch done from inside the Quiraing. On St Kilda gales blew most of the time and low cloud covered the top of the island for all but four days preventing more landscape sketching. Even when the sun did occasionally shine it caused its problems. On a visit to the Bass Rock in the Firth of Forth I began drawing a pristine white bird, a gannet, on white paper under the glaring sun. It was a disaster, as I discovered later that evening when I saw that I had quite wrongly coloured the icy-blue eyes, yellow head and shaded parts. I made sure that my next visit was in slightly overcast conditions, to give a more accurate picture of the gannet's colours!

In this book of sketches and paintings of Scottish wildlife I have inevitably had to omit many birds and animals which, if I had had the space, I would have liked to include. What I hope I have done, though, is to assemble a sufficiently representative collection of sketches to satisfy the curiosity of people who know that Scotland is famed throughout the world for its wildlife but who rarely, if ever, get the chance to see it for themselves. Scotland is unique in so many ways. In such a small geographical area it has landscapes that are both wild and pastoral, rich and barren; it has green forests and rushing streams and rivers; it has majestic moors and mountains, and fierce, windswept shorelines. Secreted in this wonderful countryside are some of the finest birds and animals in the world, and I can only hope that my drawings will give some idea and flavour of the diversity and abundance of this fascinating wildlife, and of the enjoyment it gives me as an artist and observer.

Acknowledgements

I would like to acknowledge the debt of thanks I owe to the people who have helped me in various ways throughout the year, without whom I would have had to spend many more hours looking for some species. Most are mentioned on the appropriate page but I would expressly like to thank here Allan Allison, Steve Cooper, Bert Hardie, Dr M. P. Harris, Nick Picozzi, Stuart and Robert Rae, Alf Robertson and Gordon Wright.

Schedule I Warning

Some of the birds I have drawn, such as the various birds of prey, the dotterel and the crossbill, are specially protected species during the breeding season under Schedule I of the Protection of Birds Acts 1954–67. They should not be disturbed in any way without the appropriate permit from the National Environment Research Council.

Scotland

A rough map outlining the areas where some of the drawings were done.

St Kilda (enlarged)

Stac an Armin
Stac Lee
Boreray
Soay
Soambir
Connachair 426m
Hirta
Mullach Bi
Village Bay
Ruaival
Dun
Levenish

St Kilda

(enlarged above)
Soay Sheep
St Kilda Mouse
Petrels
Shearwater

N

Mainland labels

Eynhallow
Yesnaby
Quiraing
Uists
Skye
Liathrach
Canna
Rhum
L. Badanloch
R. Alness
Rothiemurchus
Cairngorms
Deeside
Banchory
Mar
Braemar
Cairnwell
Prieshen Coire
Johnshaven
Loch Rannoch
Black Wood
Aberfeldy
Loch of Lowes
Craigvinean
R. Tay
Knapp, Inchture, home base for 1980
Perth
Tentsmuir
Strathmiglo
Loch Leven (wildfowl)
Forth
Isle of May (Seabirds)
Bass Rock (Gannets)

Orkney Islands

Common Seal
Skuas
Tystie
Hen Harrier
Merlin

Mountain & Moorland – stronghold of Deer, Blue Hare, Eagle, Peregrine, Ptarmigan, Dotterel

Relicts of **Old Pine** – **Lochs** Osprey, Capercaillie, Crossbill, Diver

Seashore & Islands Seabirds, Seals, Coastal Flora.

c. 50 miles

Islands and Seashore

The islands and coastline of Scotland provide some of the most breathtaking scenery found anywhere in the world. Each island has its own special character. Of the islands I visited in 1980 the St Kilda group of Hirta, Soay, Dun, Boreray, Stac Lee, Stac an Armin and Levenish was undoubtedly the most exciting, and this section of the book opens with the pictures and drawings I did while I was there.

The group lies 72 kilometres west of North Uist in the Hebrides. Fifty years ago the last members of the old St Kildan community were taken off the island. The young people were being lured to the mainland, and the island life was dying. Nowadays the only permanent inhabitants of St Kilda are a detachment of soldiers maintaining the missile-tracking radar station on top of the island, but the litter from the building construction scattered all round below the summit of Conachair detracts considerably from the beauty and character of this wild, desolate place.

The sheer cliffs of Conachair are the highest sea cliffs in Britain and those of Boreray reach 380 metres. Because of the stormy weather, I only managed to land on one other island, Dun, apart from Hirta where I was staying in the factor's house. However, I did manage to hitch a lift on a visiting yacht and circumnavigated the rest of the group, which enabled me to study closely the great iceberg-shaped pinnacle of Stac Lee with a snow storm of gannets swirling around the summit, and the great cliffs and pinnacles of Boreray rising sheer from the ocean depths. This group of islands forms the largest Atlantic gannetry anywhere with an estimated 60,000 pairs of gannets. Vast numbers of other seabirds also nest on these islands. Fulmars play the draughts that rise up from the great grey cliffs of the Cambir, the stacs in between it and Soay far below. The purring calls of the petrels come from the cleits (the stone storage shelters built by the St Kildans) at night when the air is alive with the sounds of manx shearwaters, Leach's and storm petrels coming in from the sea to their burrows. Young puffins make their first venture out of their burrows and head for the open sea after their parents have deserted them. Guillemots, razorbills, kittiwakes and other gulls wheel endlessly overhead.

The long isolation of some of the wildlife on St Kilda has produced differentiated species. For instance, the St Kilda wren has evolved as a separate sub-species, being slightly larger with greyer plumage and a stouter bill than its counterpart on the mainland. The wood mouse or long-tailed field mouse of St Kilda which I drew has been isolated for so long that it can be up to double the size of those on the mainland, with a completely different coat colour and behaviour pattern as well. The unique Soay sheep are the most primitive sheep in Europe and have been kept pure as a breed through many years of isolation on the island. I managed to get some drawings of rams in Village Bay where they are more accustomed to man than the sheep elsewhere on the island.

On the lovely island of Canna, the garden of the Hebrides, I again experienced the eerie shrieking calls of the manx shearwaters coming in at night to their burrows on the grassy slopes below the cliffs. I drew some head details from one of the birds which we caught during the night by torchlight and added more details from some of the burrows we studied in daylight when the pairs were roosting there prior to egg-laying. On Canna I also saw some sea eagles swoop in from the nearby island of Rhum, from where the Nature Conservancy is attempting to reintroduce this magnificent bird.

How different are the softer, more rolling islands of most of Orkney. On the small island of Eynhallow, I spent a night camping amongst the prolific seabirds; common seals lolled on the beach below with their pups. Here I sketched those neat little auks, the tystie or black guillemots, sitting on the rocks and whistling to each other with their plaintive calls. Some were feeding chicks hidden inside the old dry-stone dykes above the beach. The confiding fulmars, seemingly nesting in every nook and cranny, made excellent subjects for drawing as long as I kept out of spitting range of the contents of their stomachs! On mainland Orkney I sketched the piratical skuas, the bonxie and the Arctic skua, on the moorland above the cliffs from where they mounted their depredations on the surrounding sea-bird colonies. My attention was drawn, too, by the plants abounding in this habitat, from the lovely Scottish primrose on the headlands of Yesnaby to the brilliant blue flowers of the oyster-plant growing on the stony beaches further south.

From Orkney on to those jewels in the Firth of Forth, the Isle of May and the Bass Rock. I thoroughly enjoyed drawing the docile, approachable eider ducks sitting on their eggs amongst a carpet of white sea campion and pink thrift on the Isle of May – at least they kept still! Glossy green shags on their nests, most still with remnants of their crests, hurled abuse at me, grunting and hissing all the time. I also had to contend with the bustling, clamouring guillemots packed in row upon row on the ledges looking across to the Bass Rock, a volcanic plug standing out on the skyline, capped white with gannets.

My sketches in this section are but a few examples of the wealth of wildlife island habitats around Scotland's shores. Very few people are lucky enough to see all this coastal and island wildlife, so I hope my sketches and drawings will indicate to a wider audience something of the remarkable riches that can be found in these areas.

Leach's Petrel Chicks

Dun, St Kilda. 16.8.80

downy chick

quick sketches before the boat
came to pick me up.

more well developed
chick

both these chicks were from the
burrows outlined below on either
side of the boulder, burrows 20-24" long

3

St Kilda, Hirta

sketched on various nights
between 7th & 11th August 1980.

Leach's Petrel
⅟₁

pink gape

Storm Petrel, head & bill details

drawn from birds lured into mist
nets by tape recordings for ringing
at night. High winds prevented us
from catching as many as we hoped for

⅟₁

⁴/₃

tail and rump
of Leach's Petrel c⅟₁ amount of white feathers
very variable.

Storm Petrel, tail & rump ⅟₁

4

Storm Petrel, detail sketches from one found dead

upperside ¹/₁

underside ¹/₁

brood
patch

Hirta, St Kilda, 13–15ᵗʰ August 80

wing - 160mm
bill - 17mm
tarsus - 24.5mm

Leach's Petrel *Oceanodroma leucorhoa* ¹/₁

4ᵗʰ December 1980

plumage details from a dead bird given to
me by Dr Mike Harris, I.T.E, Banchory

plumage, upperparts blackish brown, pale brown wing coverts, white tail coverts.
underparts brown, lesser underwing coverts more rufous.

Soay sketched from Conachair,
the Cambir inbetween   22nd Aug 1980

Stac Biorach & Soay Stac
between Soay & Hirta, St Kilda   14ᵗʰ Aug 1980
drawn from a ledge part of the way down the Cambir,
the low cloud, which seems to cover Kilda most of
the time, obscuring Soay from c.600 feet upwards.

The air was full of birds, Fulmars
Gannets, Kittiwakes, Puffins
Bonxies, 2 Hooded Crows. 3
young St Kilda Wrens newly
fledged were continually calling
for food around me.

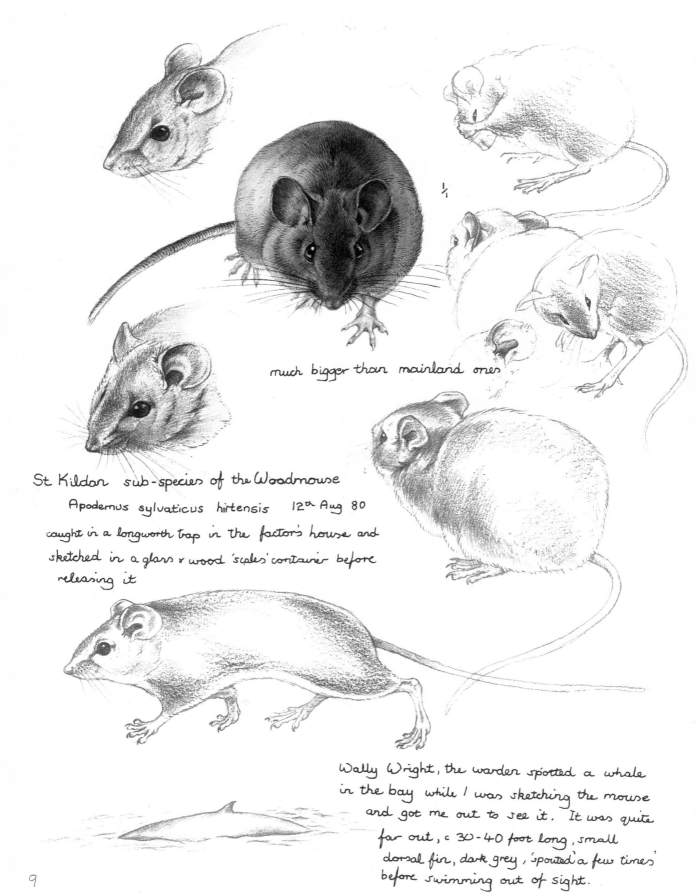

much bigger than mainland ones

St Kildan sub-species of the Woodmouse
Apodemus sylvaticus hirtensis   12th Aug 80
caught in a longworth trap in the factor's house and
sketched in a glass & wood 'scales' container before
releasing it

Wally Wright, the warden spotted a whale
in the bay while I was sketching the mouse
and got me out to see it. It was quite
far out, c 30-40 foot long, small
dorsal fin, dark grey, 'spouted a few times'
before swimming out of sight.

$\frac{1}{1}$

9

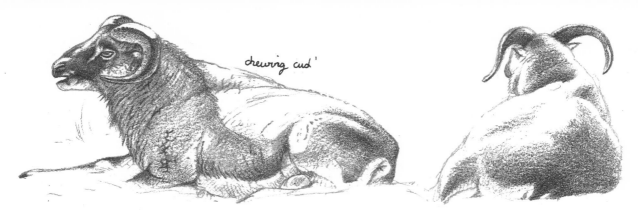

'chewing cud'

Soay Rams, Village Bay, St Kilda
8th August 1980

dark phase ram, some
more rufous, some blacker
plus the 'gingery' pale phase
as well.
very variable

youngster

L - R, Stac an Armin, Stac Lee
    and Boreray drawn from the
        slopes of Conachair 22nd Aug 80
        These islands hold the world's largest Atlantic Gannetry

11

summit Bioda Mor
576'

Dun, drawn from Ruaival, St Kilda. 16.8.80

60-80,000 pairs of Puffin breed on Dun, S. Rae
and I were across yesterday to look for rings amongst
the Puffin corpses killed by Greater B.B. Gulls for
Mike Harris (32 pairs of Gulls kill c 2,500 Puffins a year!)
We also located some Leaches Petrel chicks to see
what stage they were at.

12

St Kilda   9th Aug 1980

many still with down
on their necks.

Young Puffin ¹/₁
head & bill details

This bird was killed when it hit
a wall in the army camp. The young
puffins (just leaving their burrows for the
first time at night) are attracted to the
camp by the lights and noise of the
generators. We collect them up and
ring and weigh them before releasing
them away from the lights & noise,
511 between the 6th & 19th of August.

gape, showing
serrations on upper mandible
to grip slippery fish

underside

upperside

Manx Shearwater, St Kilda 9th Aug 1980
drawn from a bird which hit
the radio mast in the army
camp at right

74

The Village, Village Bay, Hirta, St Kilda

23ʳᵈ Aug 1980

the remains of the houses evacuated in 1930, the ones with roofs on
have been 'done up' by the National Trust for Scotland who own the
island, the cleit in the foreground conveniently obscures the
unsightly army camp.

15

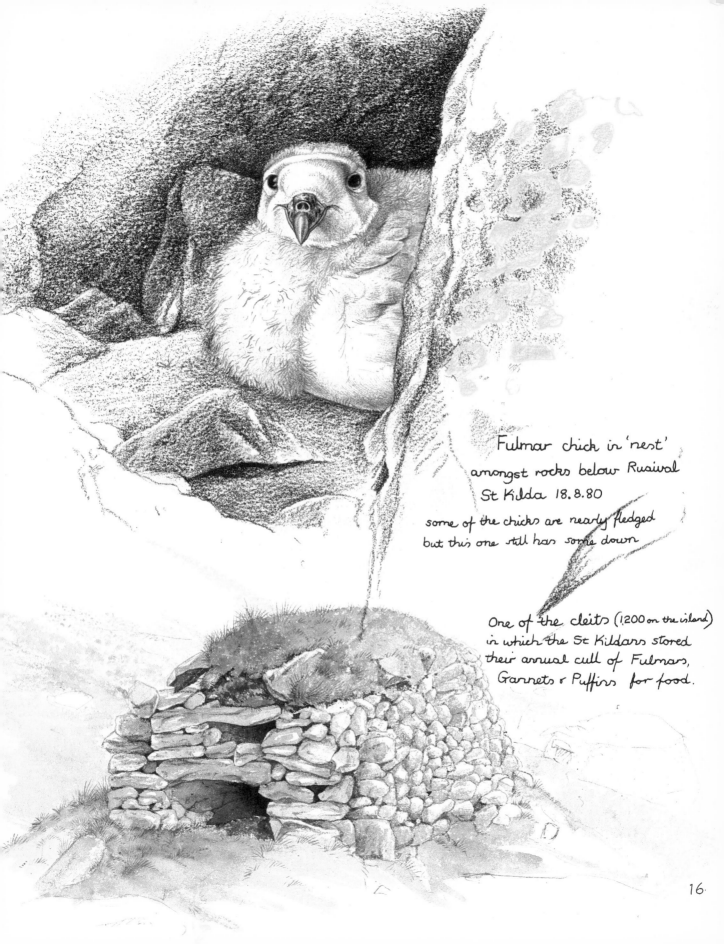

Fulmar chick in 'nest'
amongst rocks below Rusival
St Kilda 18.8.80

some of the chicks are nearly fledged
but this one still has some down

One of the cleits (1,200 on the island)
in which the St Kildans stored
their annual cull of Fulmars,
Gannets & Puffins for food.

16

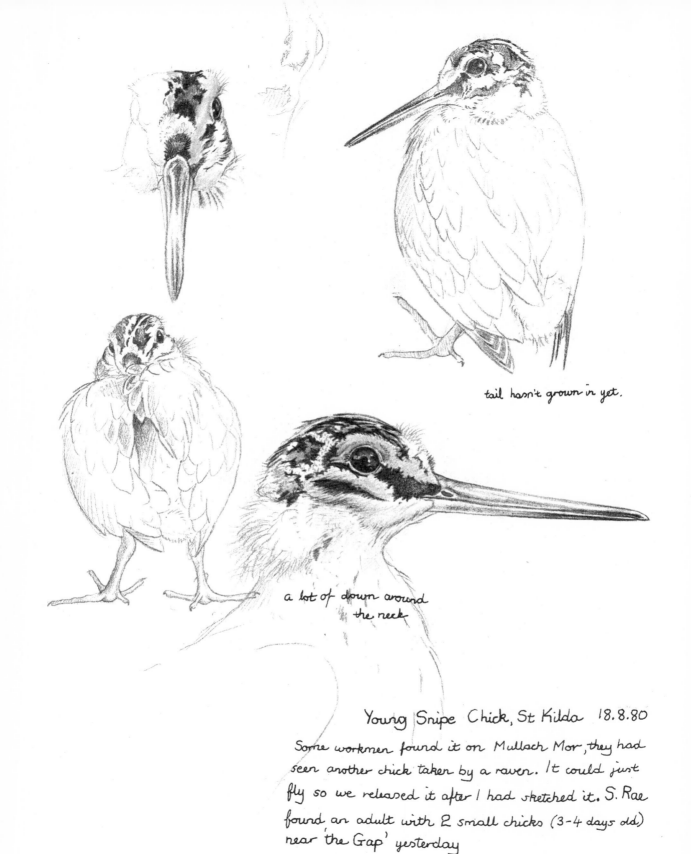

tail hasn't grown in yet.

a lot of down around
the neck

Young Snipe Chick, St Kilda 18.8.80
Some workmen found it on Mullach Mor, they had
seen another chick taken by a raven. It could just
fly so we released it after I had sketched it. S. Rae
found an adult with 2 small chicks (3-4 days old)
near 'the Gap' yesterday

Bonxie, Orkney 16th June 80

pale adult incubating 2 eggs,
plumage always looks very scruffy

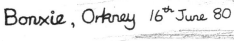

the other adult was darker

Drawn from a hide, unfortunately, this was
the only position I could draw it in as it always
faced into the strong wind

I watched this pair chasing Greater Black Backed Gulls on several
occasions making them regurgitate fish which the skuas then
caught in mid air. They are large, bulky birds with a mean
looking bill & head.

I was lucky to get some plants in flower, most had only seed heads showing as they were in between flowerings, the few in flower were very short

Scottish Primrose   Primula scotica
Orkney 24th June 80, growing in the short turf on coastal headlands, sketched below

Brough of Bigging

19

Some Coastal Plants from Orkney
June 1980

Spring Squill
Scilla verna.
growing in profusion on the short
turf above the cliffs at Yesnaby

Sea Milkwort ¹/₁
Glaux maritina

Oyster Plant ¹/₁
Mertensia maritina
growing on the shingle
along parts of the coast.

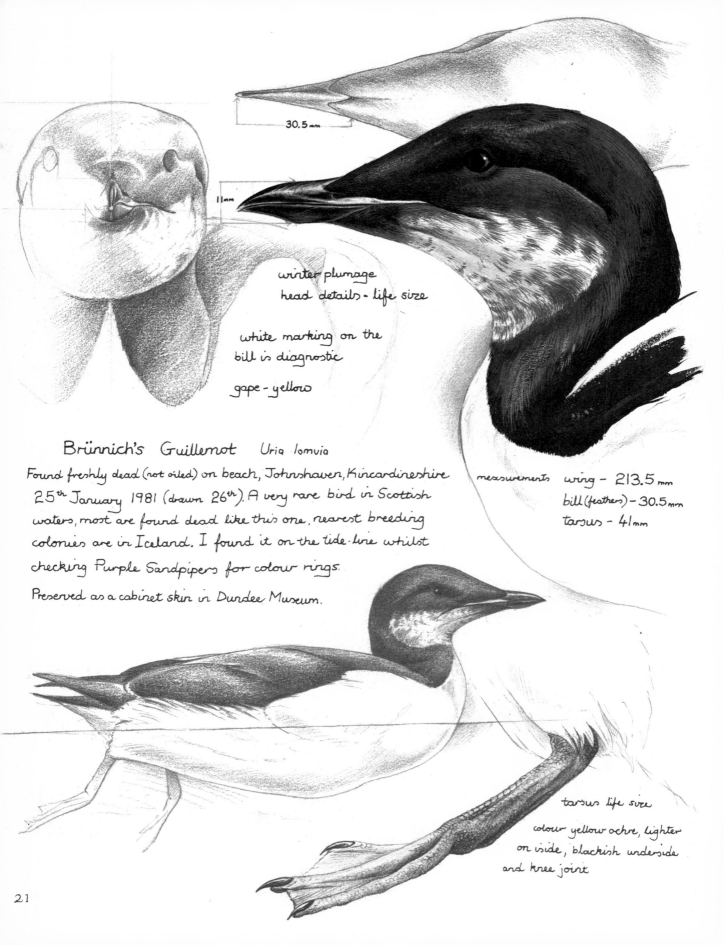

30.5 mm

11mm

winter plumage
head details - life size

white marking on the
bill is diagnostic

gape - yellow

Brünnich's Guillemot   Uria lomvia

Found freshly dead (not oiled) on beach, Johnshaven, Kincardineshire 25th January 1981 (drawn 26th). A very rare bird in Scottish waters, most are found dead like this one, nearest breeding colonies are in Iceland. I found it on the tide-line whilst checking Purple Sandpipers for colour rings.

Preserved as a cabinet skin in Dundee Museum.

measurements   wing - 213.5 mm
bill (feathers) - 30.5 mm
tarsus - 41mm

tarsus life size

colour yellow ochre, lighter
on inside, blackish underside
and knee joint

Arctic Skuas, Mainland, Orkney

I needed a hide more for protection than for camouflage at their nest. This very aggresive pair were diving at Nick Picozzi and I, stiking us on the head and landing on my back as I bent down.

. They had two very small young, one of which was still being brooded whilbt the other chick wandered around. The adults regurgitated large pieces of fish and held them in their bills while the chicks pecked at it.

chicks covered in dark brown down

dark phase adult and a very light pale phase adult made up the pair

dark bird above had a lovely silky plumage.

the pale phase adult lacked the breast band which most light birds have

cottongrass in background

22

Nest No. 267

Fulmars on nest
Eynhallow, Orkney
June 80

No. 106

very tame, they don't spit if approached slowly

Tystie (or Black Guillemot)

sketched from birds sitting around on
Eynhallow, Orkney   June 1980

drawn looking through
a telescope.

Common Seals
Eynhallow, Orkney
18ᵗʰ June 80

25

details of head & beak × $\frac{5}{4}$
showing the tubular nostrils

Manx Shearwater, drawn from
a live bird caught at night by
torchlight for ringing on Canna
with Bob Swann, 13th April 80

Around a 1,000 pairs of these
birds nest in burrows on the grassy slopes
on the cliffs shown below. The island of Rhum in the background has
about 70,000 nesting pairs

ad with chick

Gannets - Bass Rock 31st July

this drawing almost ended up in the
sea when it blew off the board.
Luckily it landed on bare rock instead
of in the mud .

Gannets, Bass Rock, 31.7.80

ground in gannetry was like a quagmire

White-tailed Eagle or Erne,

details for future ref from an adult which died at Kincraig Wildlife Park a few years ago. I borrowed it from taxidermist Allan Allison for a while to sketch it. Huge expanse of wing

underside

These eagles are being re-introduced to Scotland from the island of Rhum by the Nature Conservancy after their extermination by man. Hopefully these magnificent birds will breed next year, earlier on this year I watched four immature birds soaring together on another island.

strong talons 2/3

upperside
⅕ life size

massive hooked beak ¹⁄₁

4ᵗʰ, 5ᵗʰ October 1980

20.30hrs, 7th December 1980, Mike Nicoll, Graham Wren and I started putting up some mist nets to try and catch some waders coming in to roost by the goose pool on Tentsmuir by the mouth of the river Eden in Fife. At long last the wind had dropped enough for us to put up the line of nets, it took the three of us nearly an hour to put up and guy four nets totalling 160 feet in length. We were hampered by freezing hands and the whole area being covered with a sheet of ice which was very hazardous in the dark. Normally we would have put up more nets but this was not advisable due to the freezing weather conditions and lack of manpower to extract and process birds.

strong guy ropes with angle iron pegs 2½ feet into the sand

line of mist nets, each 9 feet high, 40 feet long, fine mesh makes them nearly invisible in the dark. each net divided into 3 by 4 shelf strings forming 'bags' which the birds get caught in (hopefully)

Within a few minutes of finishing putting up the nets we had caught 5 Oystercatchers and it was still 5 hours till high tide at 03.00 hours.

22.30 - 23.00 hours - caught 2 more Oystercatchers, built a fire with driftwood from the high tide-line to keep ourselves warm.

23.00 - 23.30 hrs. Graham has started to take photographs by flashlight of Mike and I at work with extracting and processing the waders. We caught 2 more Oystercatchers, a lot more birds are moving about now

01.00 hrs, things very busy just now constantly busy with birds, Dunlin are coming in now with small parties of Bar-tailed Godwits, Redshank and the odd Grey Plover calling.

02.00 hrs. after extracting the birds in the net we took them down before the water under the nets got too high at high tide.

The next few hours we spent processing the waders, ageing them, taking measurements of wing length, bill length, tarsus length and their weight before releasing them. We usually catch a foreign ringed each time at this site but no luck this time. Last time out we caught a Dunlin which had been rung in Finland

c × 3/2

eye detail
sketched from the grey plover held for a few minutes, detail finished off later (grey-pink gape)

The Grey Plover was the bird of the night, whoever wrote 'a bird in the hand is worth two in the bush' certainly knew what they were talking about. It is indeed a privilege to be able to handle such beautiful birds, the large eye, subtle grey gradations of plumage and the shock of the black axillaries under the wing combine to make it a most enhancing bird. Plovers all seem to have large eyes in comparison with other waders.

Final tally for the evening = 50 Dunlin (11 adult)
                              12 Oystercatcher (2 adult)
                              10 Redshank
                               3 Bar-tailed Godwit
                               1 Grey Plover   = 76 waders

05.45, finally home to a nice warm bed!

Isle of May   June 1980

Thrift
Armeria maritima

Sea Campion
Silene maritima

♀ Eider Duck on nest.
Isle of May 31.5.80

In Sea Campion next to a wall, drawn
from a metre away whilst she sat
motionless, I wish all birds were as
easy.

Eider down and a piece of eggshell
from a nest raided by gulls

34

Isle of May. 3ʳᵈ June 80

Herring Gull chicks, one still chipping
out of the egg using its 'egg tooth' on
the tip of the bill

Lesser Black Backed
Gulls
on nest

35

Guillemots
Isle of May   2nd June 80

yellow gape.

on nesting ledges, some on
very precarious positions

'bridled' form in strong
sunlight, usually darker

Shags - Isle of May   1.6.80

Forests, Lochs and Rivers

This heading covers a vast range of habitats which would really take a whole book to describe, but for my purposes it concerns the wildlife that I have observed and drawn in the Caledonian pine remnants, in the pine plantations and birchwoods, and in the various lochs and rivers throughout Scotland.

There are regrettably all too few remnants of the old Caledonian pinewoods – for example, in such places as Speyside and Deeside – for the grandeur and diverse wildlife of these areas is so rich compared with the modern monoculture forestry practised today. I especially love drawing the marvellous shapes of the old windblown pines, part of the natural cycle of woodland. The twisted grain of the wood is splendidly revealed when the bark has fallen away. If left, the rotting wood provides a good habitat for insect communities and fungi.

The roe deer that abound in the Scottish woodland are without doubt my favourite animal. I was particularly lucky to be on hand when some forestry workers found a roe fawn by the edge of some woods. Lying curled up in the green grass it was incredibly difficult to spot. I hope people will be interested in the sketches I have included that show the great variation in the shape and size of antlers due to age or, sometimes, illness. Another common animal found in abundance in the pinewoods is the red squirrel, and I had fun drawing some young red squirrels which were being cared for after some children had stolen them from their drey.

The forests have their own kinds of birds as well, such as the huge grouse and the capercaillie, the male being as large as a turkey. The capercaillie perform their courtship rituals in the early hours of the morning in the months of April and May, and this is one of the best times to observe them, for normally all one sees of a capercaillie is a rear view as it erupts from a pine tree and disappears into the distance.

Another rare species is the Scottish crossbill. The crossbills resident in the old northern pinewoods are widely regarded as a separate species because of their larger beak and wings and different calls. Although 1980 was a bad year for crossbills on Deeside because of the poor cone crop, I was able to watch flocks of the lovely scarlet cocks and green hens feeding on the available cones, then building their nests and settling down with their young.

Birds of prey haunt the Scottish woods, golden eagles nest in some high pines, buzzards are spreading east across the country despite persecution. The dead juvenile goshawk I drew was from a Scottish nest. Tawny and long-eared owls regularly hunt in the woods by night.

Another of Scotland's assets is its network of unpolluted rivers and lochs. The osprey is now increasing in numbers largely due to the efforts of the Royal Society for the Protection of Birds centred at Loch Garten in Speyside. With any luck, the osprey is now here to stay. Birds like the dramatic black-throated diver nest on highland lochs. The one I drew was the unfortunate victim of a fishing hook – a tragedy, because it is a bird we can ill afford to lose. For breeding and wintering wildfowl, Loch Leven in Kinross-shire is the most important inland water in Britain. The skeins of pink-footed geese, with their evocative calls and sounds, winging in from Iceland in September herald the arrival of the Scottish winter. Thousands of pink-footed and greylag geese, ducks and whooper swans spend most of the winter roosting and feeding on the loch and surrounding farmland. Occasionally rare visitors to the area turn up such as the Bewick's swans which I drew feeding on the stubble along with the larger whooper swans. During the summer more than a thousand ducks breed on the islands of this loch, mainly on St Serf's island.

Otters are still fairly common in Scotland, especially in the north and west and on the islands, but they are private creatures and are seldom seen. My only sketches here are of a dead otter. The rivers and lochs are also famous for their fish, in particular the salmon and the brown trout which are much sought after by anglers.

Detail drawings of a Roe Buck
shot in Craigvinean, Dunkeld 25ᵗʰ June 80

38

Juniper and
Blaeberry shrub layer
forming a good undergrowth

Speyside - 8th May - Rothiemurchus Forest, base of an old
Caledonian Scots Pine, much more beautiful than the
modern image of a straight trunk and small canopy
on pines elsewhere.

3 week old Roe Deer fawn, found

curled up in long grass in a field bordering Craigvinean
Forest. The only movement was it's twitching nose and the
flanks as it breathed, I left it as I found it.
    Dunkeld, 25th June 80

Clouds of flies around me made drawing difficult

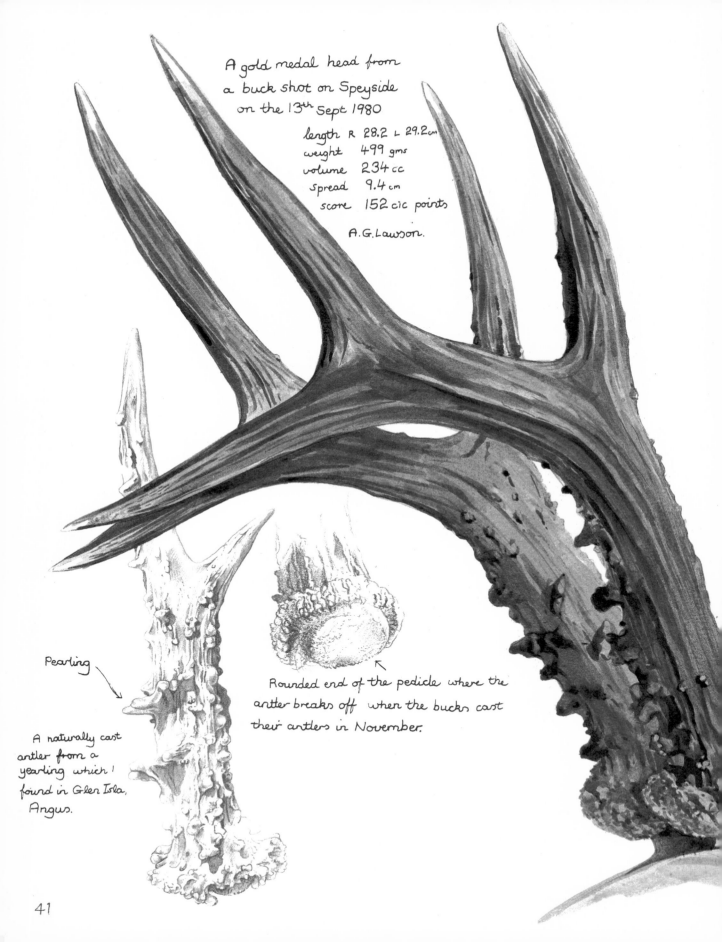

A gold medal head from
a buck shot on Speyside
on the 13th Sept 1980

length R 28.2 L 29.2cm
weight 499 gms
volume 234 cc
spread 9.4 cm
score 152 cic points

A.G. Lawson.

Pearling

A naturally cast
antler from a
yearling which I
found in Glen Isla,
Angus.

Rounded end of the pedicle where the
antler breaks off when the bucks cast
their antlers in November.

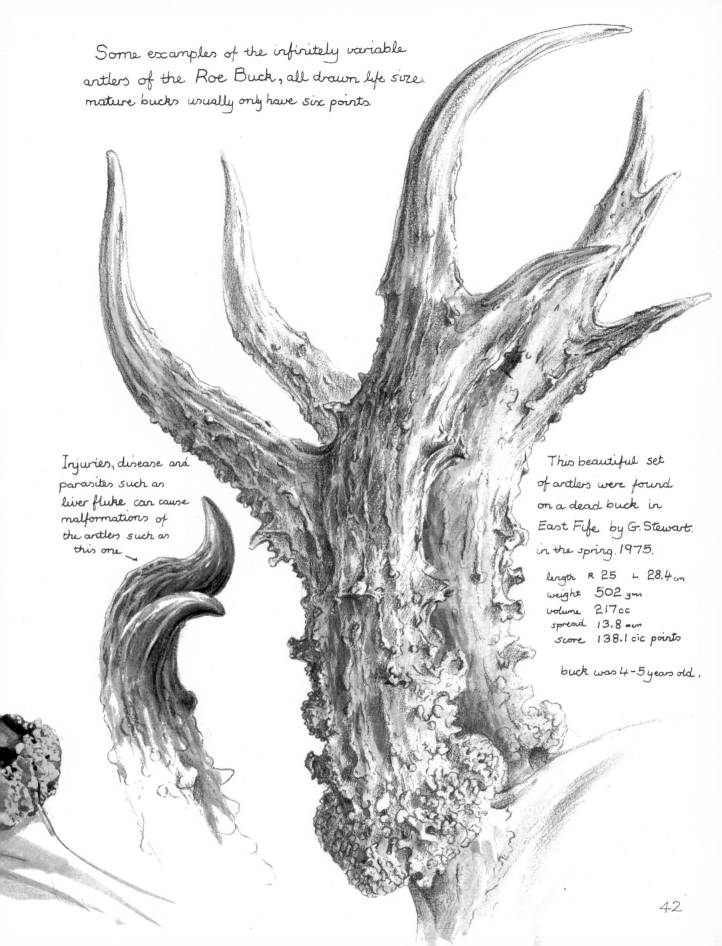

Some examples of the infinitely variable antlers of the Roe Buck, all drawn life size. mature bucks usually only have six points.

Injuries, disease and parasites such as liver fluke can cause malformations of the antlers such as this one →

This beautiful set of antlers were found on a dead buck in East Fife by G. Stewart. in the spring. 1975.

length   R 25   L 28.4 cm
weight   502 gms
volume   217 cc
spread   13.8 cm
score   138.1 cic points

buck was 4-5 years old.

♂ Red Squirrel *Sciurus vulgaris*   13-15th January 81

Detail sketches from a road casualty, Knapp road, Rossie, Perthshire, winter pelage.

cm   1   2"
5

rearfeet

43

life size

forefeet

44

45

Red Squirrel sketches
Camperdown Park Zoo, Dundee   16th January 81

46

all 'legs and head' at
this stage

climbing up a chair

Young Red Squirrel, drawn from
some (3) young squirrels being looked after
by Alf Robertson. Some kids were caught
with them after they had stolen them
from a drey

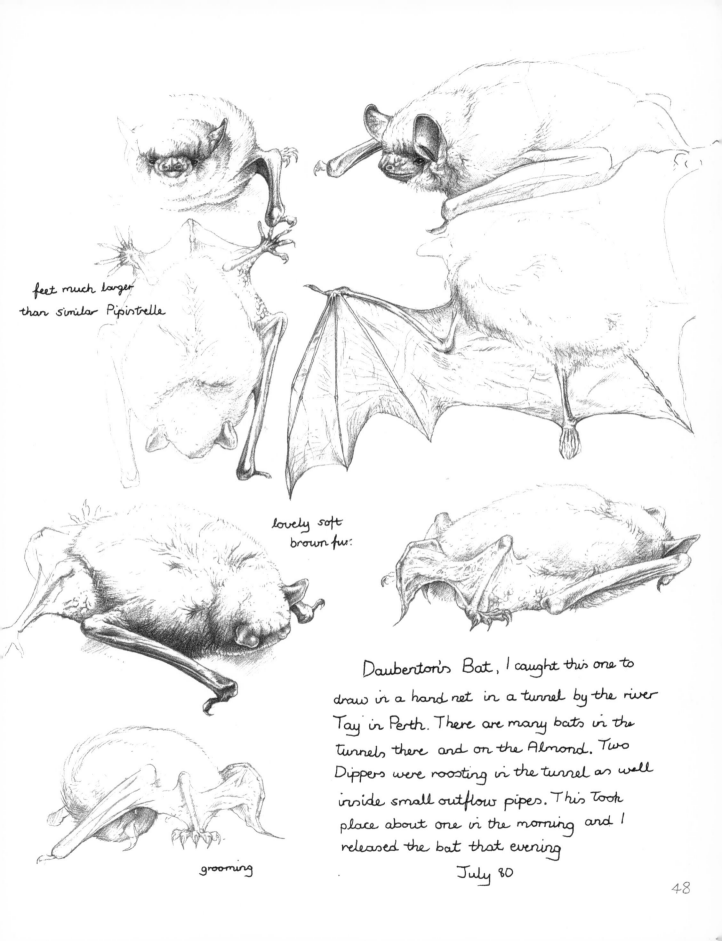

feet much larger
than similar Pipistrelle

lovely soft
brown fur.

grooming

Dauberton's Bat, I caught this one to
draw in a hand net in a tunnel by the river
Tay in Perth. There are many bats in the
tunnels there and on the Almond. Two
Dippers were roosting in the tunnel as well
inside small outflow pipes. This took
place about one in the morning and I
released the bat that evening
July 80

Otter - reference sketches from a dead one which Allan Allison was skinning, he got it from the Institute of Terrestrial Ecology, Banchory but I don't know where it originated from.

I'll never forget one sighting of an otter years ago by Little Loch Broom, Wester Ross. Whilst fishing with a hand line at high tide the weight got caught in the rocks, later that evening I went down to the shore as the tide was receding to retrieve it. I was lying on a rock with my arms in the water when an otter jumped out of the water onto the rock only a yard from me. We both lay transfixed for a while until it realized its mistake!

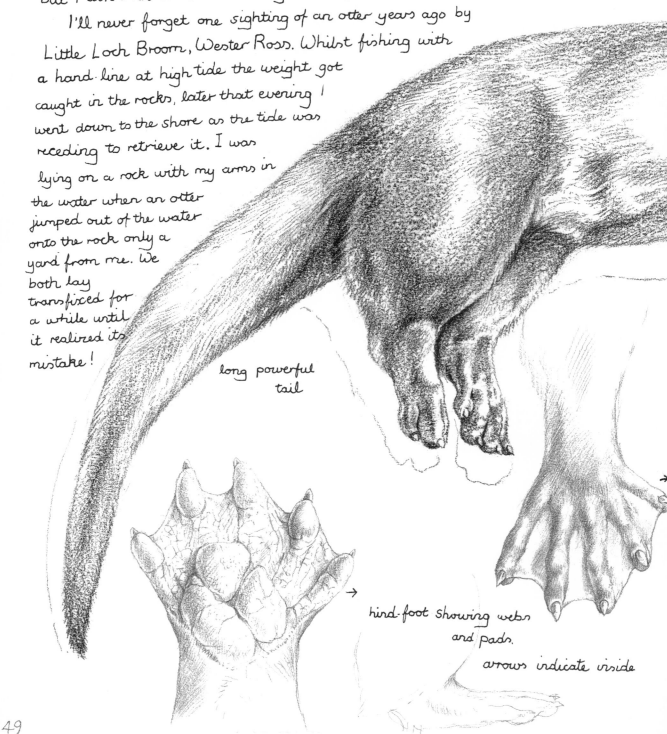

long powerful tail

hind foot showing webs and pads.

arrows indicate inside

49

fur dark chocolate brown except
for lighter throat

slightly smaller
than ½ scale.

lots of light coloured
whiskers

sketches of fore-feet showing
webs and pads

50

ancient windblown Caledonian
Scots Pine

near Mar Lodge, Deeside 17.5.80

nearly all the bark was off

51

Stark Old Caledonian Pine
Mar Forest.

Scots Pine growing out of the bank
of Allt Camghowran near Loch Rannoch
July 80

piece of pine bark
with lichen ¹⁄₁

53

Scots Pine cones eaten (split for the seed)
by Crossbills, they are scattered
on the ground underneath the trees in
which the Crossbills have been feeding.

Remains of cones eaten by
Red Squirrels

Plants and litter on the ground in the
Caledonian Pine Forest.

Cowberry, lichen, moss, birch leaves, cones & needles

Deeside, March 1980

Wood Sorrel   Oxalis acetosella.

This plant is growing in the leaf litter in most of the heron colonies I have visited so far. Most of the clover shaped leaves were hanging down as is shown in my sketch, I have also seen it growing on the open hillsides

I've put rings on over a hundred Heron chicks so far this year with more to come. They are very early this year with some birds having laid in the second week of February (early for Scotland) although some of the hill colonies were badly affected by snow in March with a foot of snow on some nests which the birds have since relaid in.

34.057

Fish Tag found in a Heron pellet. The pellets also contain small rodent remains, duck chick down etc etc, we record most of the food remains which are regurgitated by the chicks such as fish species, duck & waterhen chicks water voles etc etc.

hatched eggshell, the adults throw them to the ground below the nest

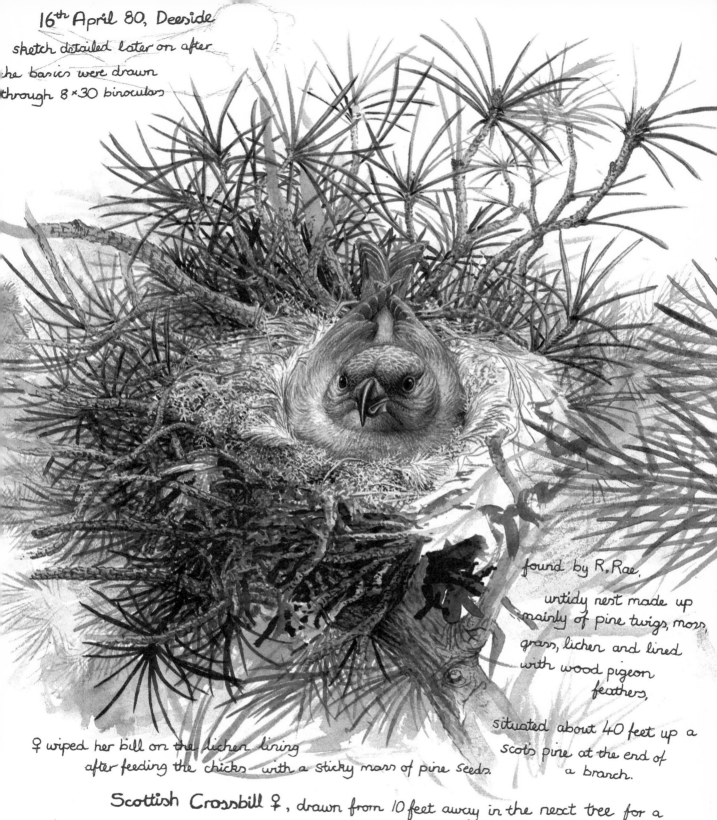

16th April 80, Deeside
sketch detailed later on after
the basics were drawn
through 8×30 binoculars

found by R. Rae.

untidy nest made up
mainly of pine twigs, moss
grass, lichen and lined
with wood pigeon
feathers,

situated about 40 feet up a
scots pine at the end of
a branch.

♀ wiped her bill on the lichen lining
after feeding the chicks with a sticky mass of pine seeds.

Scottish Crossbill ♀, drawn from 10 feet away in the next tree for a
few hours in a very uncomfortable position. They are very tame and confiding birds, the
lovely red ♂ came 3 times to feed the ♀ on the nest and she in turn fed the 4 two
day old chicks. The ♂ fed them once and even brooding them for 5 minutes while
the ♀ was away, the ♀ quivered her wings excitedly whilst he fed her.

♀ Capercaillie, I found this bird
literally wandering in circles on the Cairn
o' Mount road between Banchory & Fettercairn.
It tried to fly up into a tree but came crashing
down, it had no apparent injury but appeared
to have no balance on one side. I took it
home to draw, it died the next day
however - Aug 1980

huge powerful
bill

farred tail

Cock Caper's 'Beard' held erect

neck feathers fluffed
out in display

♂ Capercaillie, drawn from captive birds
at Blackhall, Banchory, 8ᵗʰ March, 1980

The males display at 'Leks' in April, strutting about with their
farred tails held erect and their heads held high

58

large ear hidden
behind the facial ruff.
used to locate prey at night

feathered feet with
needle sharp claws

Long Eared Owl, 6.1.1980, ⅔ life size.
found near Loch Leven, Kinross with a broken leg,
unfortunately it died after a few days

60

sketches from a tame Tawny Owl
Inchture 14th-15th Sept 80

preening

alert!

stretching

scratching

62

Tawny Owl, portraits & ref drawings

16th & 17th Sept 80

Owl from Cliff Christie, released successfully back to wild.

63

needle sharp talons

64

at rest, feathers often
fluffed out more than
this

exercising

head detail

'piercing' eyes

× 3/2

Buzzard   Buteo buteo

Sketches and plumage detail
from an injured bird from near
Aberfeldy, Perthshire

April 1980

per W & P. Mattingley

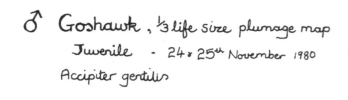

♂ Goshawk , ⅓ life size plumage map
Juvenile - 24 & 25ᵗʰ November 1980
Accipiter gentilis

upperside

¹/₁ talon

beautiful tail
feathers

→

Chances to draw a fresh
specimen of this species are very rare
so I took the opportunity of doing a
reference painting. This bird had been handed
into Dr Mike Harris at Bachory, Mick Marquiss
picked it up and gave it to me once he had
measured it. He had rung this hawk as a
chick this year and a later post-mortem will
hopefully establish the cause of death. This
rare breeding bird of prey would be much more
common if it was not so persecuted by
Gamekeepers, Egg Collectors, Birdwatchers etc etc.

513ᵐᵐ

underside

560ᵐᵐ

bill proportions c ½

wing - 328 ᵐᵐ
tail - 255 ᵐᵐ
tarsus - 79 ᵐᵐ
bill(cere) - 22.5 ᵐᵐ

68

remnant of comb from a wasp's nest (¹⁄₁)
found on the roots of a windblown pine
6ᵗʰ Nov 80

1st November 80, I found this wood pigeon chick wandering about in my goose-pen. It must be one of the ugliest chicks of all birds even at this stage of development. After sketching it, I found the nest it had fallen out of and replaced it beside the other chick. The lower mandible was much wider than the upper one

Still a lot of yellowish down on the feathers

Pair of toads ready to spawn
Morton Lochs, Fife   April 80

Oystercatcher Chicks (few days old)
23ᵈ May, by river bank Shee Water
by Cray, Glen Shee.

white

the ♂ visited the nest 6 times in 5 hours, 3
times with sticks to build up the nest and 3 times
to copulate, the nest has a very deep cup and
usually only the ♀'s white head was visible.

nest on a Scots Pine, birch
trees all around.

Osprey on nest, 1st May 1980
drawn from hide at Loch of Lowes, Perthshire
a Scottish Wildlife Trust reserve.

The ♀ on the nest had laid at least one egg, this huge nest
'framed' by birch has been used for a few years. I used a 60x
telescope but the heat haze made things difficult

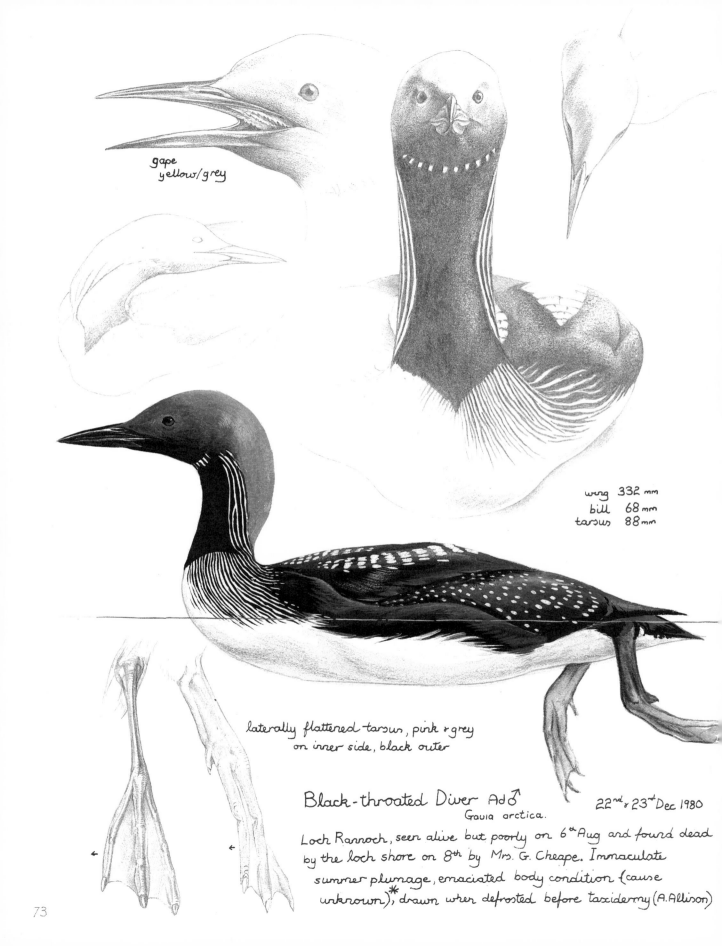

gape
yellow/grey

wing 332 mm
bill 68 mm
tarsus 88 mm

laterally flattened tarsus, pink & grey
on inner side, black outer

Black-throated Diver Ad ♂   22nd & 23rd Dec 1980
            Gavia arctica.
Loch Rannoch, seen alive but poorly on 6th Aug and found dead
by the loch shore on 8th by Mrs. G. Cheape. Immaculate
summer plumage, emaciated body condition (cause
unknown)* ; drawn when defrosted before taxidermy (A. Allison)

body & wing details c ¼ life size

* on skinning the diver Allan found a fish hook and gut stuck in and
puncturing its oesophagus next to the wishbone together with some fish remains

74

bill colour
fades very quickly

× 4/3

Wigeon Chick, 1-2 days old

Glen Esk, Angus

I found it by rapids on the river, it was very
weak and died later, I couldn't find the adult
anywhere

one doesn't normally see head detail as close

extent of colour on crown and
green around eye varies a lot.

♂ Wigeon , 1ˢᵗ December 1980
Sketches for head detail from an
adult ♂ found injured on Loch Leven
by warden Gordon Wright. Gordon
gives me some of these injured birds to
keep as they are excellent subjects for
drawing. I also picked up 2 Greylag
Geese and a Pink-footed Goose today.
Wigeon, along with Teal, are my
favourite ducks.

76

wing stretching

head scratching.

77

Pink Footed Geese

Drawings of some 'pricked' birds which
I got from G.A.Wright, warden of
Loch Leven N.N.R.
    Thousands of these geese
from Iceland winter in the
Scottish lowlands, especially
around Loch Leven

serrations on bill for
cropping grass

adult, young birds are duller and less well marked
with 'mottled' undersides.

sleeping

alert

head a bit
too small.

Greylag Goose
Anser anser

More robust, longer legs and larger bill
than the Pink-footed Goose

79

wing stretching

feeding

These sketches are from injured geese, they had been 'winged' by wildfowlers and had been picked up on Loch Leven by the Nature Conservancy warden G.A.Wright. I keep them in a small collection for drawing from.

80

Loch Leven 13ᵗʰ Nov 80,

I did these sketches whilst waiting for a cannon net to fire, two nets had been set the night before on a grassy point hopefully to catch some of the 400 odd Wigeon feeding on the grass. After a five hour wait we had to give up as stubble burning put the Wigeon off just as they were begining to concentrate on the shore.

Two of the Mute Swans were wearing yellow colour rings, they had been rung at the Montrose Basin, Angus, in August. Over 200 of the moulting flock were caught there.

♂ Wigeon asleep

sketches of ♂ Mallards   2ⁿᵈ Oct 80
drawn from a hide on St Serfs Island
L. Leven

82

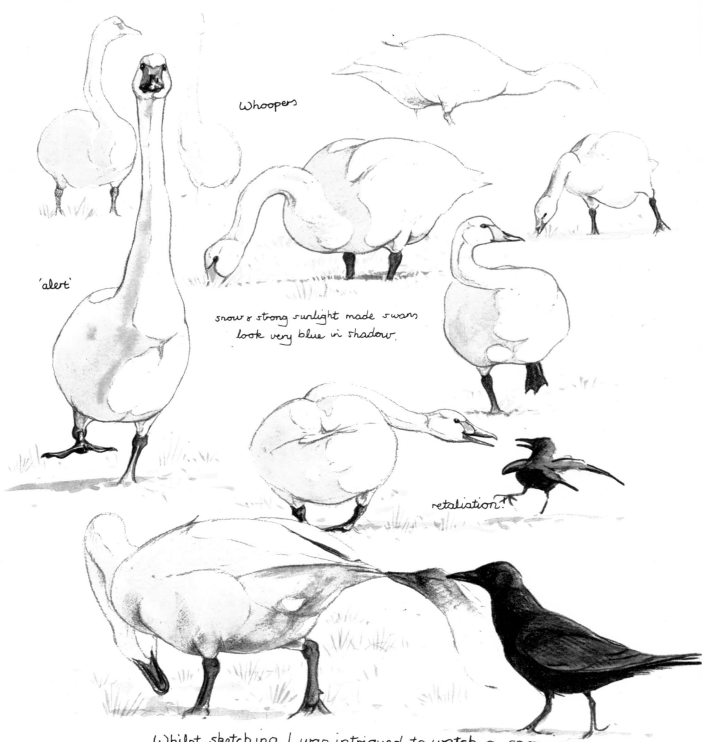

Whoopers

'alert'

snow & strong sunlight made swans
look very blue in shadow.

retaliation.

Whilst sketching I was intrigued to watch a carrion crow
striding up to a pair of Bewick's which were nearest to me. To my
astonishment it proceeded to pull the swans tail, only jumping back
when the swan lashed out angrily at it. This 'cat and mouse' game
went on for fully five minutes, the swan even had to put up with
another crow which joined in for a few pulls! I can think of no other
reason for this behaviour than sheer devilment.

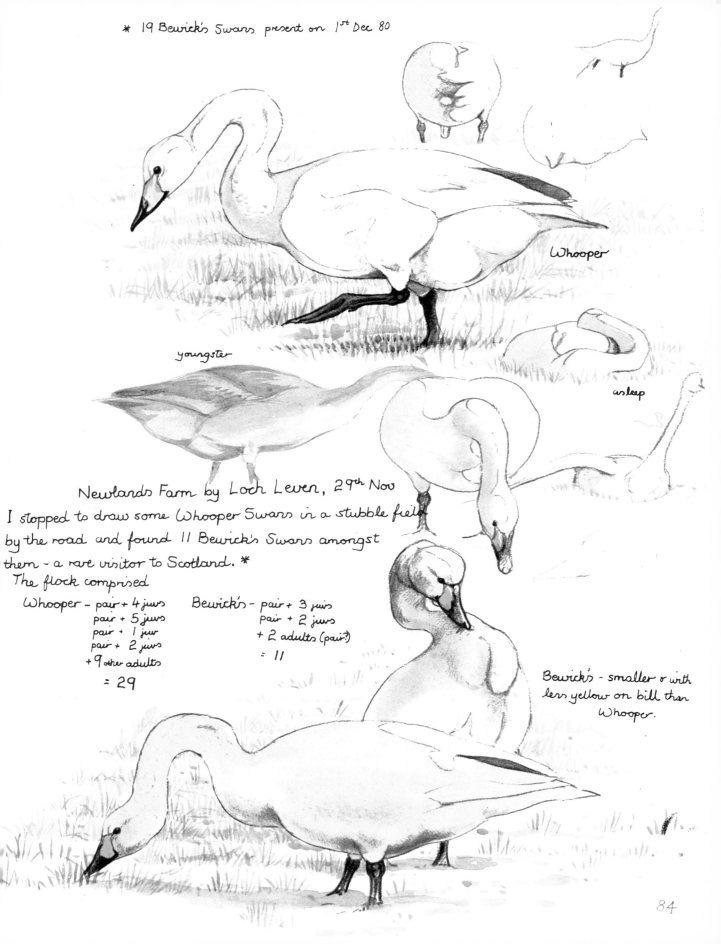

* 19 Bewick's Swans present on 1st Dec 80

Whooper

youngster

asleep

Newlands Farm by Loch Leven, 29th Nov

I stopped to draw some Whooper Swans in a stubble field by the road and found 11 Bewick's Swans amongst them - a rare visitor to Scotland. *

The flock comprised

Whooper - pair + 4 juvs
pair + 5 juvs
pair + 1 juv
pair + 2 juvs
+ 9 other adults
= 29

Bewick's - pair + 3 juvs
pair + 2 juvs
+ 2 adults (pair?)
= 11

Bewick's - smaller & with less yellow on bill than Whooper.

84

Whooper Swans, Loch Leven, 30th October 80
too cold to draw for long due to the hard frost, the
'haunting' trumpeting calls of the swans carrying over the still loch.

This individual, the nearest to me, had the
reflections off the ripples running up its
neck at regular intervals, a most
unusual effect
♂ Gadwall feeding behind

dozing off

loch very calm, lots of feathers on the surface
as many of the ducks are still moulting

This display I watched consisted
of 6 adults facing each other and
trumpeting loudly whilst flapping
their wings wildly and raising
and lowering their heads. This
went on for 3 minutes

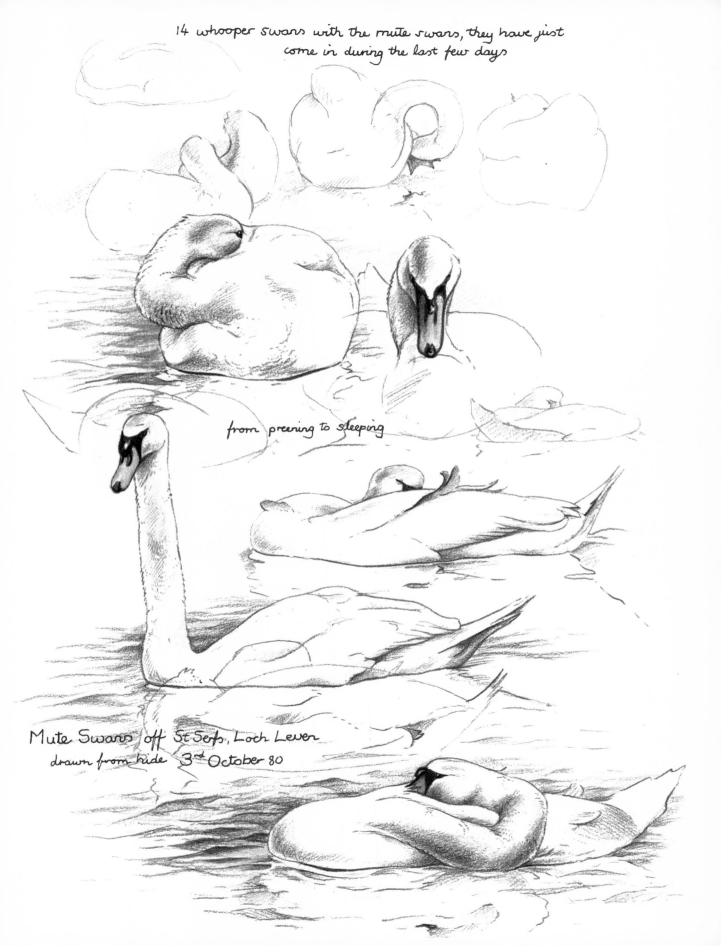

14 whooper swans with the mute swans, they have just
come in during the last few days

from preening to sleeping

Mute Swans off St Serfs, Loch Leven
drawn from hide 3rd October 80

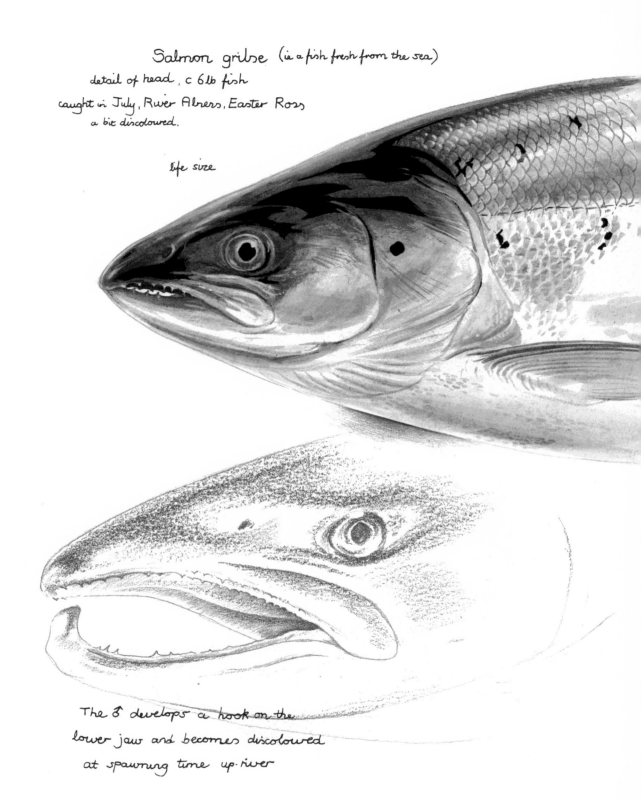

Salmon grilse (is a fish fresh from the sea)
detail of head, c 6lb fish
caught in July, River Alness, Easter Ross
a bit discoloured.

life size

The ♂ develops a hook on the
lower jaw and becomes discoloured
at spawning time up·river

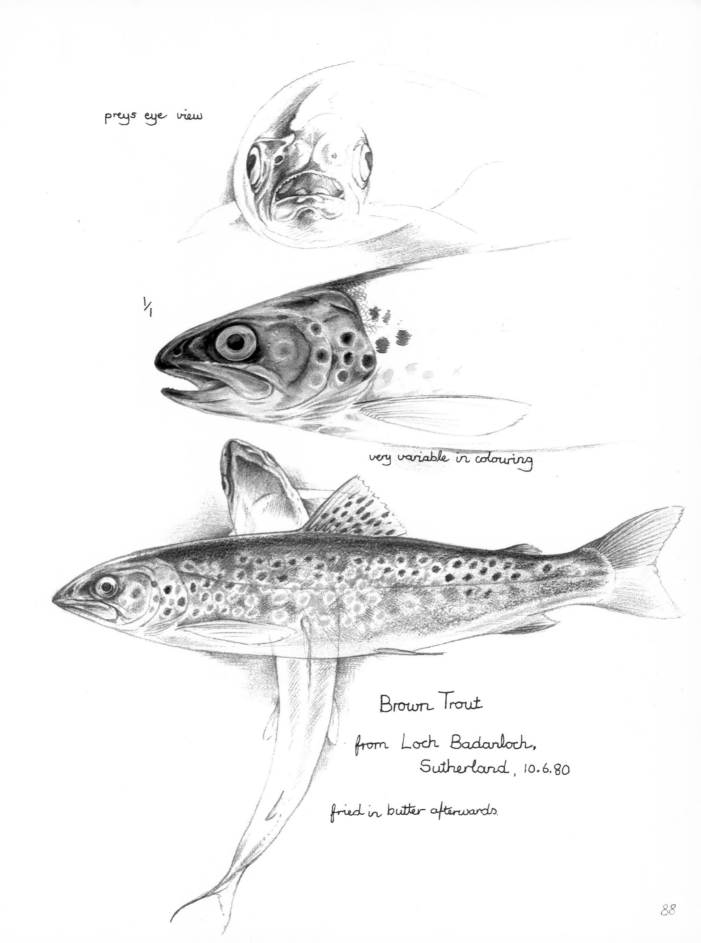

prey's eye view

$\frac{1}{1}$

very variable in colouring

Brown Trout

from Loch Badanloch,
Sutherland, 10.6.80

fried in butter afterwards.

Wing of ♂ Pintail, life size, from A. Allison.
'Speculum' more coppery in some lights

89

Mountains and Moorland

Mountain and moorland, collectively known in Scotland as 'the hill', cover more or less three-quarters of the country from the flat coastal moors of Sutherland to the high summit of Ben Nevis and the Cairngorm plateau. This great wilderness, largely unpolluted, forms a sanctuary for many forms of wildlife much threatened in the rest of Europe. For instance it is the stronghold of the golden eagle and peregrine falcon, two birds which seem to me to embody the very spirit of freedom that is found in these remote areas. It may be a biased opinion but most of the species found on the hill seem to possess a lot more character than their lowland counterparts.

Some of the species I have drawn are also found in the other two habitat sections of this book. Roe deer are just as likely to be seen on the open moorland far from the woodland with which they are normally associated. Peregrine falcons are found on the offshore islands and coastline. Frogs are found on lower wetlands, too, but I suspect they are now much more common on the pool-studded moorland; on some days the moorland literally seems to be hopping with frogs. Plants like the wood anemone are found up in the corries, a relic of the days when the woodland once reached high up onto the hill.

On lower moorland such as in Orkney I was able to draw the hen harrier where it still breeds in abundance. I had to spend ten hours in the hide because, unfortunately for me, the female harrier was away hunting most of the time as the male was not bringing in any food. At least I was kept amused by the antics of the reptilian chicks squabbling amongst themselves. Watching and drawing a female merlin on her nest was a new experience, and nearby I sketched a kestrel chick from a nest in the heather. On a Sutherland moor I sketched the old stump and roots of an ancient pine uncovered from the peat by a burn in spate. Here I also drew the heath-spotted orchid and butterwort which grow in profusion in the wetter areas. Higher up at least three pairs of greenshank had young and the adults constantly 'chipped' their warning calls. I watched five greenshank mobbing a female hen harrier and almost driving her into the loch.

In the lower glens of Perthshire and Angus I sketched some of the breeding waders such as the lapwing, but sadly the range of waders such as the redshank and curlew is contracting because of the effects of drainage and forestry ploughing. Further up these glens the golden eagle and peregrine falcon nest on the high cliffs and in the corries. I keep a check on some of these pairs, marking and ringing the chicks and collecting data on things like their food remains. Sights like a peregrine 'joy-flying' above its nesting cliff in March, performing incredible aerobatics against a backdrop of snow and wind-curled icicles, are the sort of rewards one gets for one's patience.

The open heather moorland above the corries is the domain of the golden plover, red grouse and the red deer herds. I spent quite a while out with some stalkers learning about the red deer and sketching and experiencing the stalking at first hand. Apart from sheep farming, deer stalking and grouse shooting are still the main land uses of the moorland.

Higher up onto the summits of the hills the moorland gives way to the tundra-type vegetation of the tops, the haunt of the dotterel, ptarmigan and mountain hare. The dotterel is one of the real treasures of the tops, and it would be rather a disappointment if it had to nest in potato fields at sea level, as dotterels do in the Netherlands. The mountain hare, too, has a special place in my affection. I sketched it in its winter and autumn coats, as well as sketching leverets at very tiny and at half grown stages. These little animals have a tough life with all the elements against them.

Then there are the mountain flora, the lovely splashes of colour that brighten up a harsh environment such as the harebell or Scottish bluebell springing out of the rock scree, or the vivid pinkish flowers of the rare alpine catchfly growing in Angus, their only site in Scotland. Clumps of yellow mountain saxifrage grow alongside mountain streams and on wet cliffs. Dazzling patches of freshly flowered moss campion contrast starkly with the grey rock below an eagle eyrie. With such an array of wild flowers, in addition to the fauna of the area, it is hardly surprising that I find mountain and moorland so appealing.

'The Needle'
Quiraing, Skye, 6.8.80

90

'flat out'

Red Deer, drawn at a deer farm near
Auchtermuchty, courtesy of Dr Fletcher.
16th July 80

The heads are quite awkward
to draw, the skin on the stags antlers
appears to 'stretch' the rest of the face.
Older deer appear to have
lighter eyes.

most in their sleek summer coats

in 'velvet'

hind 'chewing cud'

92

Taking a stag off the hill

Setting off to stalk the stags after sighting them

shooting the stag from up to 300 metres with a high velocity rifle (243 or 270 calibre.

Gralloching the stag,
ie bleeding and disembowling
the beast after shooting it

Taking the stag off the hill down to
the landrover for transportation to the larder where
it is cleaned further and awaits collection by
the game dealer.

crown

tres - tine (or tray)

beam

brow tine

(bay) bez - tine

Royal Stag shot by
Robert Wittmann with stalker
Bert Hardie just below Meall
Tionail, Byrack, Mar Estate

It was a heavy stag and the three
of us had to drag it over the hill to
the landdrover, 15th October 1980

95

details of a stag shot
by a German Client with
stalker Bert Hardie on
Carn Bhac, Mar Estate
24th Sept 80

96

18th November 1980, I was out with stalkers Bert Hardie and Lovat Fraser and ghillie Stuart Rae who are culling the hinds just now. Between November and January they will cull 250-300 hinds out of a population of between 3-4,000 on the estate. Today we set off up the Corriemulzie Burn and up through the saddle between Carn Mór and Carn na Drochaide and worked this area into Glen Ey. There were plenty of deer in the area but most were in the wrong areas for stalking with the wind. It was very cold with fresh snow from the 550 metre contour upwards. The lines of the deer tracks made beautiful weaving patterns visible from a long way off. By the end of the day we had shot 6 hinds and one calf. If a hind has a calf when shot they usually try and shoot the calf as well as they are sometimes still fairly dependant on the mother.

sketches from the head of one of the shot hinds.

Most of the Mountain Hares are white now, we also watched a pair of Eagles soaring over Carn na Drochaide for a while

Mar Forest, Braemar 25th Sept, sketch of a gralloched stag which I had just dragged off the hill, it was soaked after Thérèse helped me drag it through a burn to the landrover.

On Mar Estate up to 100 stags and up to 300 hinds are culled every year, roughly 10-15% of the deer on the ground. With stags only the worst heads are shot, ie those with malformed antlers (such as the one sketched above with double brow tines), switches and hummels. The stags are mostly shot by clients, mainly Germans, the stalker chooses which beast the client is to shoot.

Red Deer Stag roaring during the Rut, coming in to wallow in a peat hag. 1980

Carn Liath, Mar Forest, Braemar.
26th Sept 80

'Foally', led by
Thérèse, the pony ghillie

drawn from the back of the landrover which she was tied to,
waiting to bring the stag off the hill when it had been shot.
She was looking very dejected standing in the wind and rain, note
the special saddle and harness for the job.

huge hinds feet

Blue or Mountain Hare,
which turns white in the winter
for camouflage.

fluffed out cheeks

Hare from Craigvinean, Dunkeld
Feb. 1980
caught live for me by N. Drummond
and later released.

102

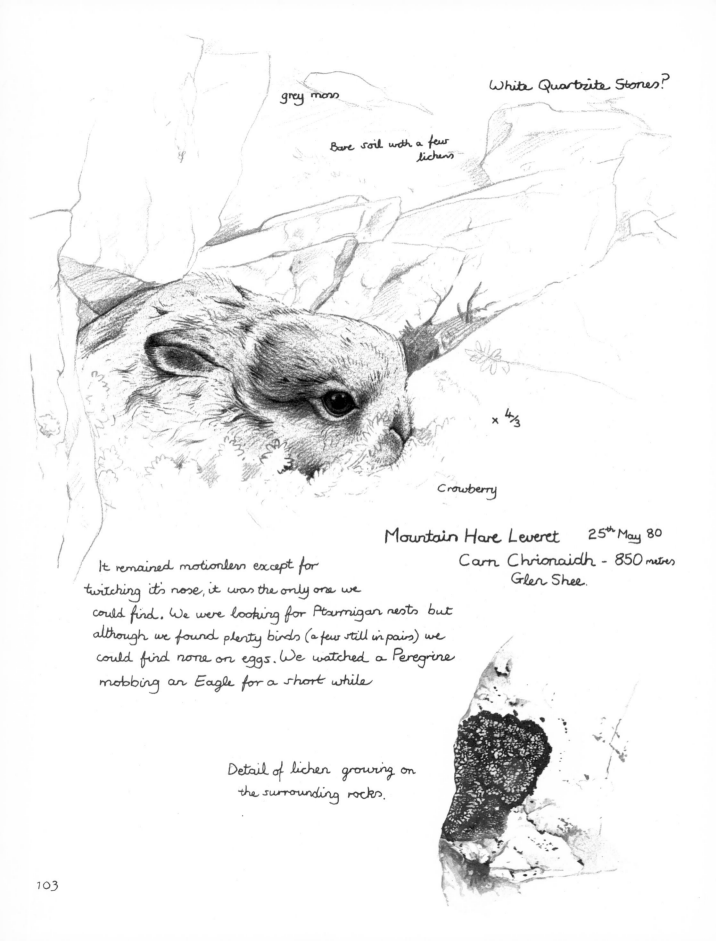

grey moss

White Quartzite Stones?

Bare soil with a few lichens

× 4/3

Crowberry

Mountain Hare Leveret   25th May 80
Carn Chrionaidh - 850 metres
Glen Shee.

It remained motionless except for twitching it's nose, it was the only one we could find. We were looking for Ptarmigan nests but although we found plenty birds (a few still in pairs) we could find none on eggs. We watched a Peregrine mobbing an Eagle for a short while

Detail of lichen growing on the surrounding rocks.

103

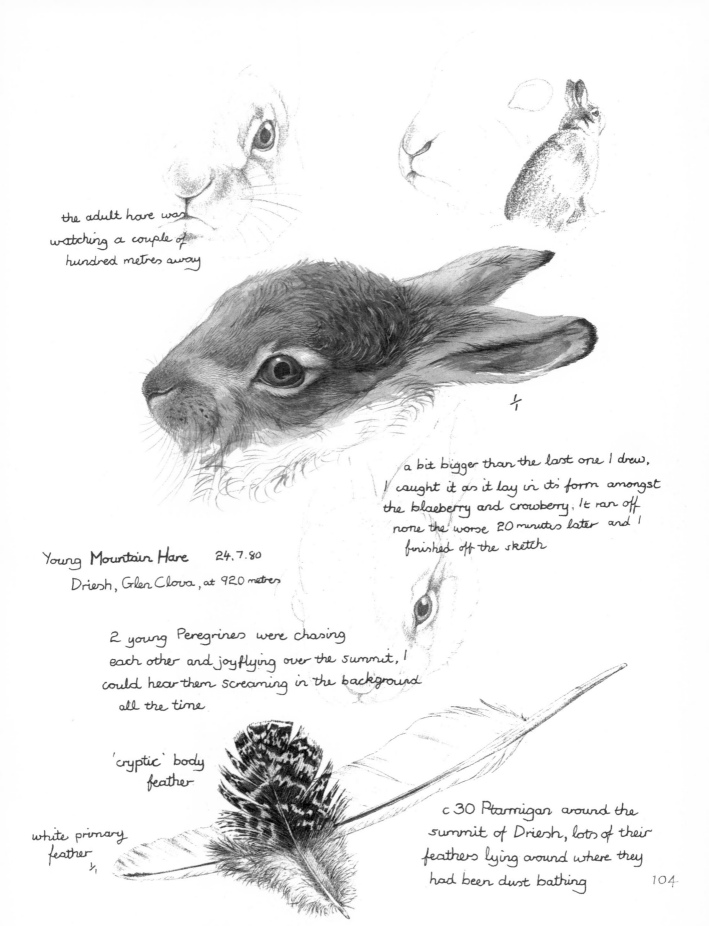

the adult hare was
watching a couple of
hundred metres away

$\frac{1}{1}$

a bit bigger than the last one I drew,
I caught it as it lay in its form amongst
the blaeberry and crowberry. It ran off
none the worse 20 minutes later and I
finished off the sketch

Young Mountain Hare   24.7.80
Driesh, Glen Clova, at 920 metres

2 young Peregrines were chasing
each other and joyflying over the summit, I
could hear them screaming in the background
all the time

'cryptic' body
feather

white primary
feather
$\frac{1}{1}$

c 30 Ptarmigan around the
summit of Driesh, lots of their
feathers lying around where they
had been dust bathing

104

'running sketches' from the hare.

detail of autumn coat
with some white hair coming
through

hare's snowshoes

Mountain Hare, 14th October 1980, a road
casualty which I picked up near the Devil's Elbow,
Cairnwell, it's coat was starting to turn white for
the winter months. A fair bit of snow on the high
tops already.

Liathach (1053 metres)

Glen Torridon, Wester Ross.

Summit obscured by cloud on this wild day 23rd January 81

2 Ravens and a Buzzard flying around

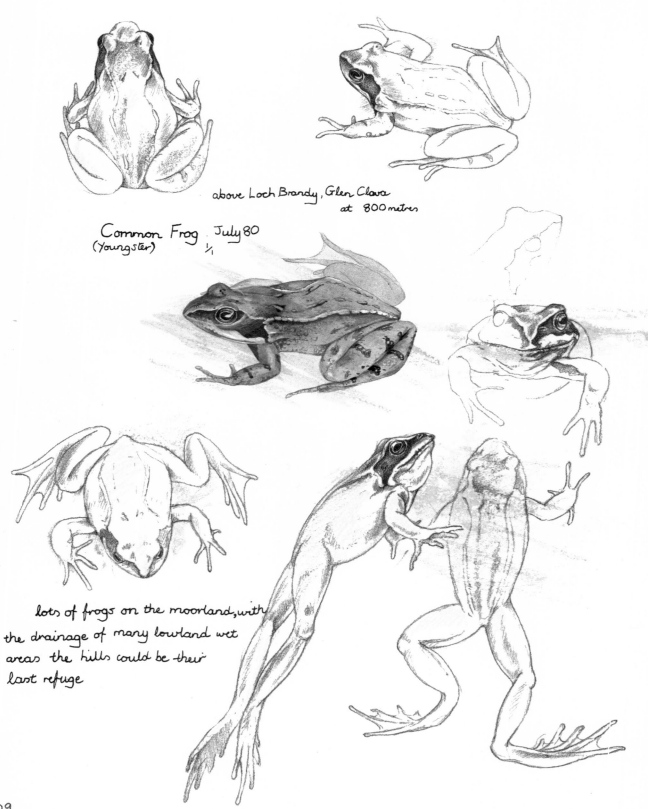

above Loch Brandy, Glen Clova
at 800 metres

Common Frog. July 80
(Youngster) 1/1

lots of frogs on the moorland, with
the drainage of many lowland wet
areas the hills could be their
last refuge

Stoat, 31.3.80, found dead
on the road, Tarland, Deeside.

Just a few traces of the white
winter coat left

distinctive black tip
to the tail present in it's ermine (winter)
coat as well.

110

Golden Eagle being mobbed
by House Martins which nest under
the overhang, Angus corrie

(comp) Dec 1980

feathers coming through
the down

nest lined with
woodrush - now flattened

Eagle chick   number Z40361, that is it's ring number, 30th May 80
c 5 weeks old. This sketch is copied from the original which got wet in
my pocket whilst climbing down from the eyrie in an Angus glen. Snow
had fallen overnight and was melting making the rock and vegetation
extremely wet and I got thoroughly soaked. The chick put on a half
hearted aggressive display before it's head flopped back, still too weak
to put up much of a show with those huge talons. Both adults present
at first but only the ♀ kept flying past at intervals till I left.

Remains of food found in the eyries I visited
        included - Mountain Hare   Lepus timidus
                Rabbit   Oryctolagus cuniculus
                Ptarmigan   Lagopus mutus
                Red Grouse   Lagopus l. scoticus
      and even a fledgling Ring Ouzel   Turdus torquatus

The ♀ at one site repeatedly stooped at me, breaking off only a couple of feet from my head!

cere - light blue with pink fusing in

eye ring - light blue

gape pink, red inside

scrape on a grassy ledge.

lots bones r feathers

23rd May 80, Angus Glen, brood of 4 chicks, sketched quickly while I was checking on the nest, there isn't time to sketch really because of disturbance and the adults screaming could attract attention from unwanted visitors.

Here are some of the prey items I recorded in the Peregrines eyries during the 1980 season

fragment of hatched eggshell from an eyrie, faded from its former beautiful colouring

Racing and Feral Pigeons
Red Grouse (adults)
Golden Plover (1 adult)
Redshank (adults)
Lapwing (adults)
Snipe (1 adult + 1 c.10 day old chick)
Purple Sandpiper (1 adult in Summer Plumage!)
Great Spotted Woodpecker (1)
Blackbirds
Mistle Thrush
Meadow Pipits

Footnote - A man from South Shields was fined £500 + costs at Sunderland Court, 3rd Nov 80 for stealing 2 Peregrines for falconry from a Perthshire eyrie, I had marked and rung the chicks earlier and prosecution was brought by the R.S.P.B.

114

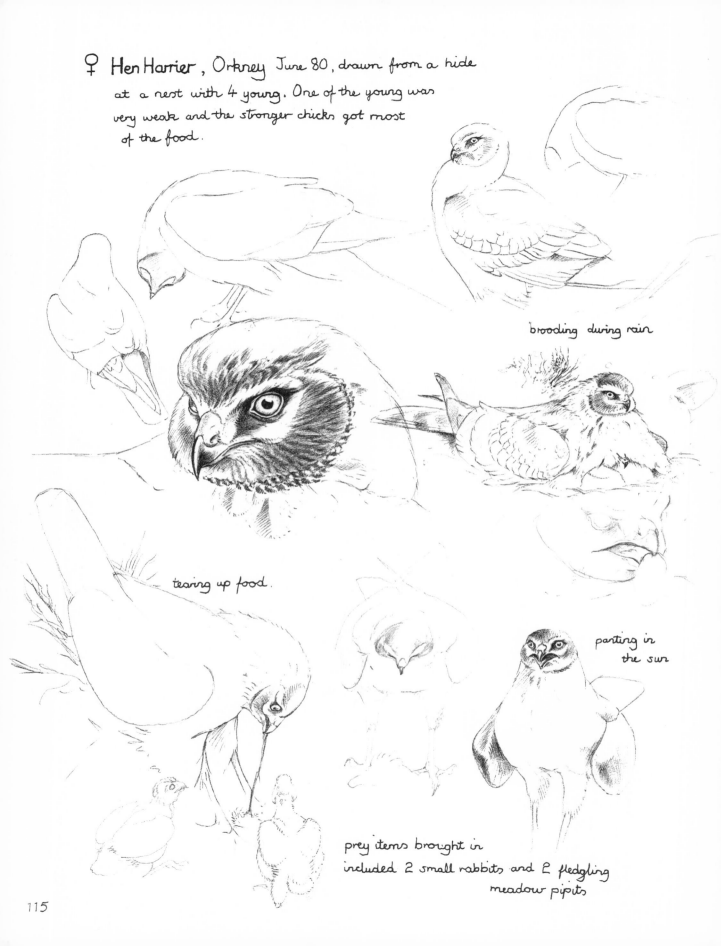

♀ Hen Harrier, Orkney June 80, drawn from a hide
at a nest with 4 young. One of the young was
very weak and the stronger chicks got most
of the food.

brooding during rain

tearing up food.

parting in
the sun

prey items brought in
included 2 small rabbits and 2 fledgling
meadow pipits

115

whilst hunting near the nest she was mobbed by a Merlin which had a nest nearby and also by a Short-Eared Owl.

she tidied the nest, taking away the prey remains and pellets and dropped them a few hundred yards away

she got very annoyed at the flies around the nest, constantly snapping at them

one of the young pipits was still alive when she brought it in

Drawn from a photographer's hide
15 feet from nest

♀ Merlin, Orkney June 80, brooding two
chicks in the rain, she gradually got wetter
and wetter. She looked very uncomfortable
and the chicks gave her a rough time
pushing about under her

two addled eggs as well.

♂ called her off twice
and she returned to the chicks with the food, young
meadow pipits on both occasions

detail sketches, ♀ Kestrel (injured)
drawn from a bird held in my hand
April 1980

This bird got its revenge by grabbing me on the nose whilst
I was drawing its head, I still have the scars!

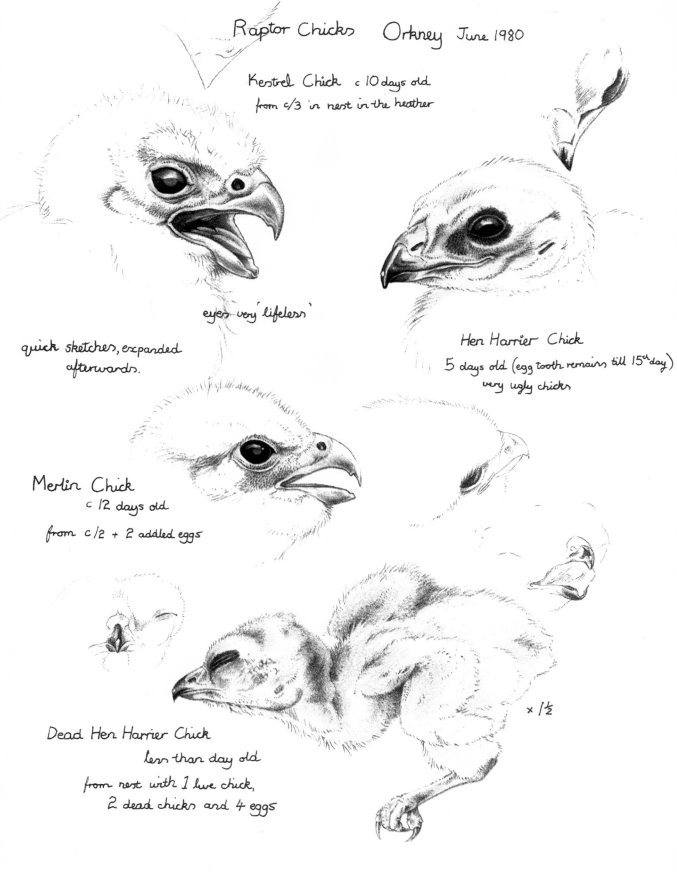

Raptor Chicks Orkney June 1980

Kestrel Chick c 10 days old
from c/3 in nest in the heather

eyes very 'lifeless'

quick sketches, expanded
afterwards.

Hen Harrier Chick
5 days old (egg tooth remains till 15th day)
very ugly chicks

Merlin Chick
c 12 days old
from c/2 + 2 addled eggs

x 1/2

Dead Hen Harrier Chick
less than day old
from nest with 1 live chick,
2 dead chicks and 4 eggs

head detail
♂ showing the red wattle
above the eye

pink gape

more dowdy ♀

Red Grouse, drawn from captive
birds at Blackhall, Banchory. Beak
details & ♀ from a dead bird
May 1980

egg ½, very variable
markings

feathered feet

120

♂ Dotterel on nest c. 3 eggs
drawn from a short distance away
with the aid of binoculars. It's eye
was very large when alarmed by
Golden Plover alarm calls.

greeny sheen
to back feathers

one scapular was very dark
brown with red/buff edging, this ♂
had a very bright head.

another ♂ Dotterel on nest, I
started to draw it and five
minutes later it got up and
led 3 chicks away, they
were not more than a day
old.

erect posture when
alarmed.

Sma' Glen, Perthshire
12.5.80

♂ on nest, crest
'flying' in the wind

♀ on nest, half asleep in the heat of the sun

Lapwing, both adults sketched while incubating, I used the car as a hide and drew them with the aid of a telescope. The ♂ has a longer crest, darker and more distinct facial markings. They kept having to chase away sheep with their lambs which frequently came close to their nest. An Oystercatcher was incubating 3 eggs close by.

small lapwing chicks, drawn from several which I was putting rings on, most are just hatching now

Golden Plover on nest

drawn for 45 minutes from a photographer's
hide 15 feet from it's nest on high moorland

There don't seem to be as many birds on the hill
this year, at least on the Angus hills compared
to what there usually is.

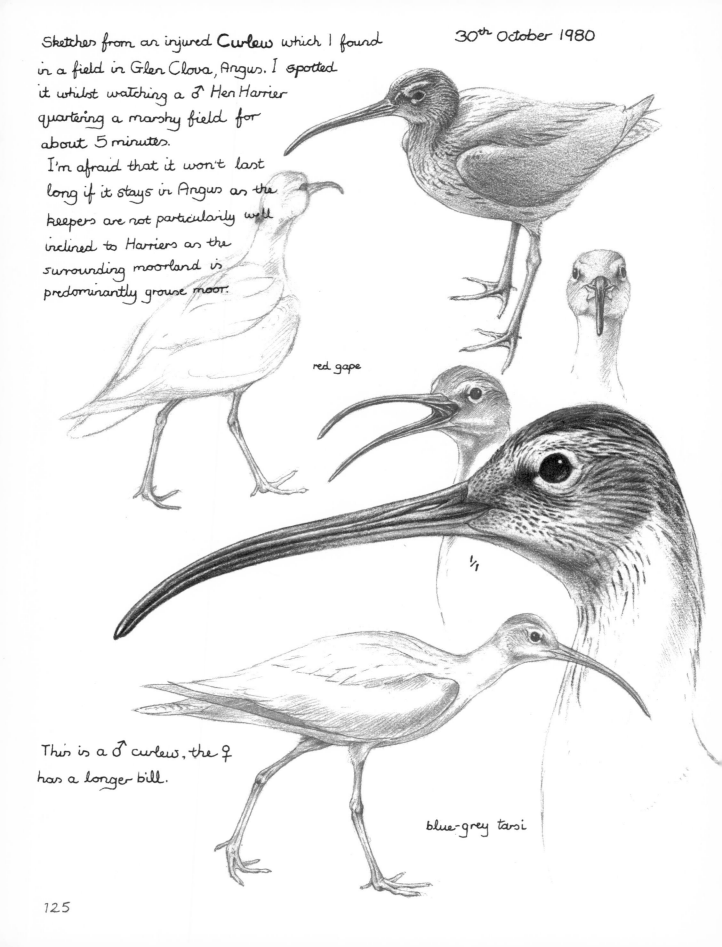

Sketches from an injured **Curlew** which I found
in a field in Glen Clova, Angus. I spotted
it whilst watching a ♂ Hen Harrier
quartering a marshy field for
about 5 minutes.
  I'm afraid that it won't last
long if it stays in Angus as the
keepers are not particularly well
inclined to Harriers as the
surrounding moorland is
predominantly grouse moor.

30ᵗʰ October 1980

red gape

¹/₁

This is a ♂ curlew, the ♀
has a longer bill.

blue-grey tarsi

125

23rd April 1980
Angus & Perthshire

many plants in
woods as well.

Wood Anemone, Anemone nemorosa
many flowers scattered on the
hillside just appearing from areas
where snow has melted. Old grass
still flattened, leaves on the
anemone curled up

I visited five Peregrine Falcon
territories and a Golden Eagle territory today to check for signs of
occupation.

Perg Site 1 - clutch 4 eggs, new higher eyrie location
  "   " 2 -   "   "   " , new location (not usual old raven nest) *
  "   " 3 -   No birds present & no signs, Kestrel on usual site, adult
                 Peregrine was shot off nest 1979!
  "   " 4 -   clutch 4 eggs, traditional nesting ledge.
  "   " 5 -   no signs of any birds at all (chicks stolen 1979)

Eagle - Watched nesting cliffs for 2 hours from a distance before seeing
        adult leaving cliff face. Twenty minutes later the adult came back
        soaring high over the corrie, suddenly she closed her wings and
        plummeted down into the corrie curving into the cliff braking with
        widespread wings only seconds before landing on the eyrie. I won't
        go near the nest till the chick/chicks hatch

* clutch stolen later that week, I found remains of a toilet roll in the nest - used
   to wrap the eggs in!

126

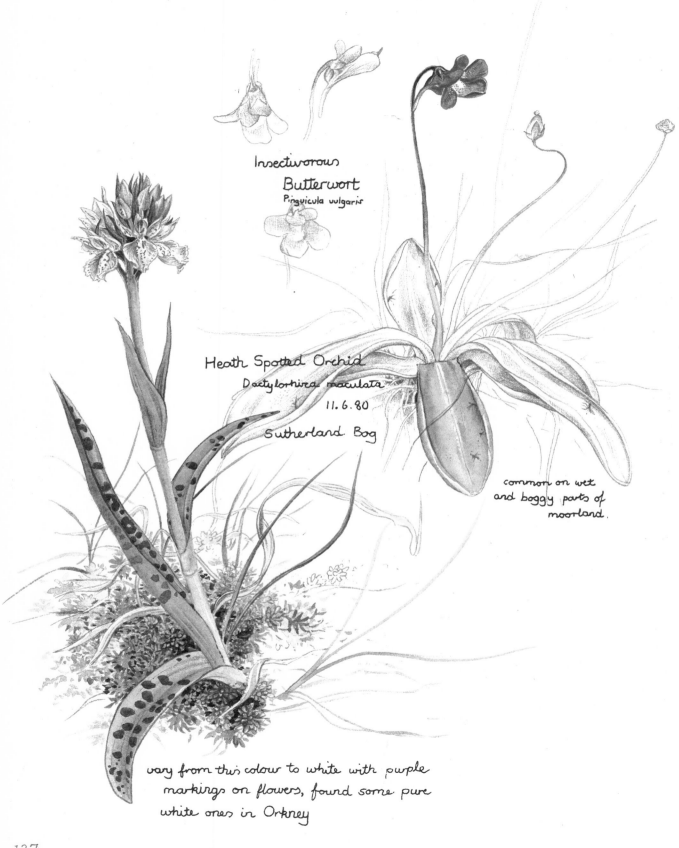

Insectivorous
Butterwort
Pinguicula vulgaris

Heath Spotted Orchid
Dactylorhiza maculata
11. 6. 80
Sutherland Bog

common on wet
and boggy parts of
moorland.

vary from this colour to white with purple
markings on flowers, found some pure
white ones in Orkney

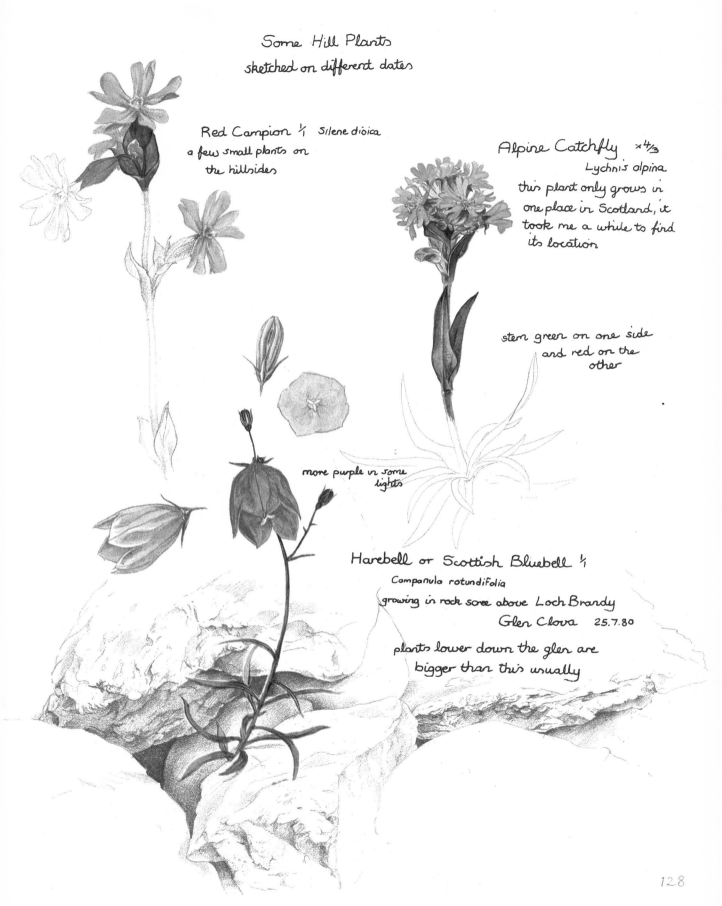

Some Hill Plants
sketched on different dates

Red Campion ⅟₁   Silene dioica
a few small plants on
the hillsides

Alpine Catchfly   ×⁴⁄₃
Lychnis alpina
this plant only grows in
one place in Scotland, it
took me a while to find
its location

stem green on one side
and red on the
other

more purple in some
lights

Harebell or Scottish Bluebell ⅟₁
Campanula rotundifolia
growing in rock scree above Loch Brandy
Glen Clova   25.7.80

plants lower down the glen are
bigger than this usually

128

Looking down on a clump of Yellow Saxifrage
growing by a burn in   *Saxifraga aizoides*
            Glearn Taitreach.
    It was growing in profusion along the stream
    banks at the head of the glearn.
                            19.7.80